Numerical and Practical Exercises in Thermoluminescence

Vasilis Pagonis   George Kitis   Claudio Furetta

Numerical and Practical Exercises in Thermoluminescence

With 110 Figures

 Springer

Vasilis Pagonis
Department of Physics
Mcdaniel College
Westminster MD 21157
USA

George Kitis
Nuclear Physics Laboratory
Aristotle University of
Thessaloniki
Thessaloniki 540 06
Greece

Claudio Furetta
Department of Physics
University of Rome,
La Sapienza
Rome 00185
Italy

Cover illustration: Typical thermoluminescence glow curves for first and second order kinetics. The equations for the glow curves are also shown. Figures are drawn by the authors.

Mathematica is a registered trademark of Wolfram Research, Inc

Library of Congress Control Number: 2005926338

ISBN-10: 0-387-26063-3   e-ISBN 0-387-30090-2
ISBN-13: 978-0387-26063-1

Printed on acid-free paper.

Printed in the United States of America.   (TB/SBA)

9 8 7 6 5 4 3 2 1

springer.com

I dedicate this book to my wife, **Mary Jo Boylan**, whose constant encouragement and support made this book possible.

Vasilis Pagonis

I dedicate this book to my wife, $\Phi\omega\tau\epsilon\iota\nu\eta$, and my mentor, **Professor Stephanos Charalambous**.

George Kitis

I am very grateful to my wife, **Maria Clotilde**, for her loving support of my scientific work.

Claudio Furetta

Preface

Thermoluminescence (TL) is a well-established technique widely used in dosimetric and dating applications.

Although several excellent reference books exist which document both the theoretical and experimental aspects of TL, there is a general lack of books that deal with specific numerical and practical aspects of analyzing TL data. Many times the practical details of analyzing numerical TL glow curves and of applying theoretical models are difficult to find in the published literature.

The purpose of this book is to provide a practical guide for both established researchers and for new graduate students entering the field of TL and is intended to be used in conjunction with and as a practical supplement of standard textbooks in the field.

Chapter 1 lays the mathematical groundwork for subsequent chapters by presenting the fundamental mathematical expressions most commonly used for analyzing experimental TL data.

Chapter 2 presents comprehensive examples of TL data analysis for glow curves following first-, second-, and general-order kinetics. Detailed analysis of numerical data is presented by using a variety of methods found in the TL literature, with particular emphasis in the practical aspects and pitfalls that researchers may encounter. Special emphasis is placed on the need to use several different methods to analyze the same TL data, as well as on the necessity to analyze glow curves obtained under different experimental conditions. Unfortunately, the literature contains many published papers that claim a specific kinetic order for a TL peak in a dosimetric material, based only on a peak shape analysis. It is hoped that the detailed examples provided in Chapter 2 will encourage more comprehensive studies of TL properties of materials, based on the simultaneous use of several different methods of analysis.

Although the subject of TL curve fitting and glow curve deconvolution is beyond the scope of this book, the readers may find the spreadsheet examples in Chapter 2 useful and easily adaptable for implementing simple curve fitting algorithms. These algorithms are based on the experimentally measurable maximum TL peak height and the corresponding temperature (I_M and T_M). In the examples given, the activation energy E acts as the adjustable parameter in the computerized curve fitting procedure. Several of these curve fitting spreadsheet examples can also be found in the authors' website.

Chapter 3 presents for the first time in the TL literature detailed numerical examples of several commonly used theoretical models, as well as several comparative studies of analytical expressions used for kinetic analysis of TL data. The main thrust of this chapter is to illustrate how to solve the differential equations describing the traffic of carriers during the various TL processes in the crystal. A few simple examples of solving the basic differential equations of TL using a spreadsheet are given mostly for illustrative and educational purposes. The main body of this chapter consists of a gradual presentation of increasingly complex TL models using the program *Mathematica*.

We have found this programming environment to be very efficient, versatile, and easy to work with, once the basic structure and programming style have been mastered. We emphasize in particular the transparent nature of the numerical integrating techniques used in *Mathematica*, which are particularly suited for solving systems of "stiff" differential equations that are common in theoretical TL work. The *Mathematica* numerical integration code is very stable and efficient, with rare occasional numerical instabilities. All examples given in this book have typical running times of 1–2 min on a desktop computer. Several of the *Mathematica* examples given in this book can also be found in the authors' website for easy reference and download.

In Chapter 4, we give numerical exercises relevant to the TL dose response of dosimetric materials. The models described in this chapter are taken directly from the published TL literature in order to facilitate direct comparison of the results with the original papers. As much as possible, we have kept the same symbols and mathematical notation as the original papers for easy cross-reference. The *Mathematica* programs are given in a "modular" form consisting of a small core of subroutines performing separate tasks, which can be easily adopted by the readers for a variety of different purposes.

A very important class of TL models is presented, namely models based on competition during irradiation process, competition during the TL heating process, as well as models containing competition during both irradiation and heating. The last exercise in Chapter 4 presents a numerical example of how the superlinearity and supralinearity coefficients $g(D)$ and $f(D)$ can be calculated from experimental TL versus dose curves.

In Chapter 5, we present a variety of exercises dealing with practical aspects of several phenomena commonly encountered in the study of TL materials. A group of four numerical exercises deal with the accuracy and reproducibility of measurements performed using TL dosimeters (TLDs). In particular, we show how the statistical accuracy and reproducibility of TL data can be greatly improved by using individual correction factors for each TLD. The next two exercises deal with the commonly observed phenomenon of thermal quenching and comprise a detailed simulation of thermal quenching effects on the measured TL glow curves and on the initial rise technique. The next group of two exercises deals with aspects of the mathematical formalism used in environmental TL dosimetry.

Two extensive exercises in Chapter 5 concern with the important but somehow underutilized technique of the TL-like presentation of phosphorescence decay curves and with the practical aspects of how to correct experimental TL data for temperature lag effects between the heating element in TL equipment and the sample itself.

Astute readers will notice the absence from this book of any exercises dealing with dating applications of TL. We decided that such exercises were beyond the specific scope of this book and refer the readers to the review papers in the annotated bibliography, as well as to the monographs dedicated to this important topic.

Perhaps, one of the most useful aspects of this book is the inclusion of an annotated bibliography on TL topics. To the best of our knowledge, there has been no other published annotated bibliography in the TL literature, and we believe that this will be an important tool for both established TL researchers and persons starting a research project in this field.

Although it is not possible to give a comprehensive annotated bibliography, we have provided characteristic examples of published articles in the various topics covered in this book. Several review articles of general interest on TL have also been included in our listings; these can serve as important introductory material for the various topics.

Our choices of papers and monographs in the annotated bibliography were dictated by our desire to guide the reader toward few characteristic and complete examples of TL data analysis, rather than providing a complete list.

An Appendix is provided with examples of the most basic commands in *Mathematica* for reference purposes, although it can only cover the most rudimentary aspects of this powerful programming environment.

November 9, 2005

Vasilis Pagonis
George Kitis
Claudio Furetta

Contents

List of Figures

List of Tables

1
Expressions for Evaluating the Kinetic Parameters

Introduction

Thermoluminescence (TL) is defined as the emission of light from a semiconductor or an insulator when it is heated, due to the previous absorption of energy from irradiation. The graph of the amount of light emitted during the TL process as a function of the sample temperature is known as a "TL glow curve."

During a thermoluminescence experiment, one typically obtains several glow curves under different conditions. For example, a series of TL glow curves may be obtained for a material that was irradiated at several different doses, or was preannealed at various temperatures. Usually the main goal of measuring and analyzing these TL glow curves is the extraction of several parameters that can be used to describe the TL process in the material. Examples of these parameters are the activation energy E for the TL traps (also called the trap depth), the frequency factor s, the order of kinetics b of the TL process, the capture cross-sections for the traps and recombination centers, and the concentrations of these traps and centers.

In this chapter the various theoretical methods and analytical expressions used to analyze TL glow curves are presented, and they serve as a reference material for the rest of the chapters in the book. For more extensive descriptions of the various models that lead to the expressions in this chapter, the reader is referred to the research articles and textbooks listed at the end of this chapter, as well as to the annotated bibliography at the end of this book.

Simple Thermoluminescence Model

The process by which materials emit light when heated can be understood by considering the simplest possible model consisting of two localized levels, an isolated electron trap (T) and a recombination center (RC), as shown in Figure 1.1. This is commonly referred to as the one-trap-one-recombination center model (OTOR).

Let us denote by N the total concentration of the traps in the crystal (m^{-3}), by $n(t)$ the concentration of filled traps in the crystal (in m^{-3}) at time t, and by $n_h(t)$

1

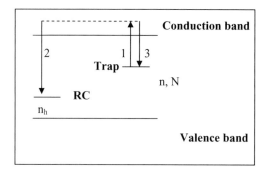

FIGURE 1.1. The simple two-level model for the thermoluminescence process.

the concentration of trapped holes in the recombination center (in m^{-3}). The initial concentration of filled traps at time $t = 0$ is denoted by n_0.

In a typical thermoluminescence experiment the sample is heated with a linear heating rate $\beta = dT/dt$ from room temperature up to a high temperature usually around 500°C. As the temperature of the sample is increased, the trapped electrons in T are thermally released into the conduction band, as shown by the arrow for transition 1 in Figure 1.1. These conduction band electrons can either recombine with holes in the recombination center RC (transition 2), or they can be retrapped into the electron trap T (transition 3), as shown in Figure 1.1. The intensity of the emitted light is equal to the rate of recombination of holes and electrons in the recombination center, and is given by

$$I(t) = -\frac{dn_h}{dt} \tag{1.1}$$

Figure 1.2 shows in the schematic diagram the increase in sample temperature T, the simultaneous emission of light $I(t)$, and the corresponding decrease in the concentration n_h of trapped holes.

Expressions for First-, Second-, and General-Order TL Kinetics

The equations governing the thermoluminescence processes have been given by Randall–Wilkins [1], Garlick–Gibson [2] and May–Partridge [3] for first, second, and general orders, respectively:

$$I(t) = -\frac{dn}{dt} = nse^{-E/kT} \tag{1.2}$$

$$I(t) = -\frac{dn}{dt} = \frac{n^2}{N}se^{-E/kT} \tag{1.3}$$

$$I(t) = -\frac{dn}{dt} = n^b s'e^{-E/kT} \tag{1.4}$$

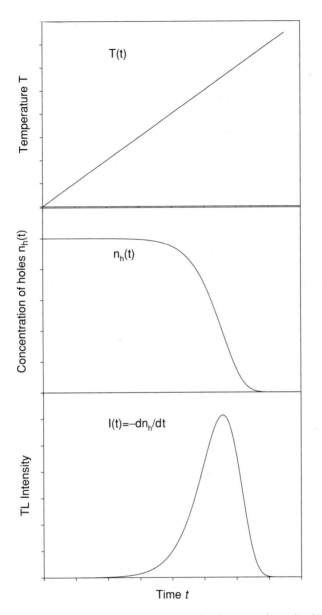

FIGURE 1.2. The temperature profile $T(t)$, the thermoluminescence intensity $I(t)$, and the concentration of trapped holes $n_h(t)$ in the recombination center RC as a function of time t. A linear heating rate β is used during the experiment.

where

$E = $ the activation energy or trap depth (eV)

$k = $ Boltzmann's constant (eV K^{-1})

$t = $ time (s)

$T = $ the absolute temperature (K)

In a typical experimental situation a linear heating rate β is used to heat the sample, resulting in the temperature varying as $T = T_0 + \beta t$, where $\beta =$ linear heating rate (K s^{-1}), and

$T_0 =$ temperature at time $t = 0$ (K)

$s\ \ =$ a constant characteristic of the electron trap, called the "preexponential frequency factor" or "attempt-to-escape frequency" (s^{-1}). This parameter is proportional to the frequency of the collisions of the electron with the lattice phonons. Typically the maximum values of s correspond to the values of the lattice vibration frequency, i.e. $10^{12} - 10^{14}$ s^{-1}.

$N\ =$ the total trap concentration (m^{-3})

$n\ \ =$ concentration of trapped electrons at time t (m^{-3})

$b\ \ =$ the kinetic order, a parameter with values typically between 1 and 2

$s'\ =$ the so-called effective preexponential factor for general order kinetics (m$^{3(b-1)}$s^{-1}).

Equations (1.2)–(1.4) can be integrated by assuming a linear heating rate β, and the following equations are obtained:

$$I(T) = n_0 s \exp\left(-\frac{E}{kT}\right) \exp\left[-\frac{s}{\beta} \int_{T_0}^{T} \exp\left(-\frac{E}{kT'}\right) dT'\right] \qquad (1.5)$$

$$I(T) = n_0^2 \frac{s}{N} \exp\left(-\frac{E}{kT}\right) \left[1 + \frac{n_0 s}{\beta N} \int_{T_0}^{T} \exp\left(-\frac{E}{kT'}\right) dT'\right]^{-2} \qquad (1.6)$$

$$I(T) = s'' n_0 \exp\left(-\frac{E}{kT}\right) \left[1 + \frac{s''(b-1)}{\beta} \int_{T_0}^{T} \exp\left(-\frac{E}{kT'}\right) dT'\right]^{-\frac{b}{b-1}} \qquad (1.7)$$

In equations (1.5)–(1.7), the additional parameters are:

$n_0 =$ number of trapped electrons at time $t = 0$ (m^{-3})

$s'' = s' n_0^{(b-1)} =$ an empirical parameter acting as an "effective" frequency factor for general-order kinetics (in s^{-1}).

It must be noted that although equations (1.2) and (1.3) can be derived from simple thermoluminescence models by using certain simplifying assumptions, the general-order kinetics equation (1.4) is completely empirical and in general will have no relationship to actual physical models.

Figure 1.3 shows a comparison of TL glow peaks for first- and second-order kinetics. In the case of second-order kinetics the emission of light is "delayed" by the retrapping of the electrons in the trap T; this delay shows mostly during the descending part of the TL glow curve.

Figure 1.4 shows several first-order and second-order TL glow peaks calculated by using equations (1.5) and (1.6), and for different initial concentrations n_0 of trapped electrons. It is noted that for TL glow peaks following first-order kinetics, the temperature T_M of maximum TL intensity does not depend on the initial

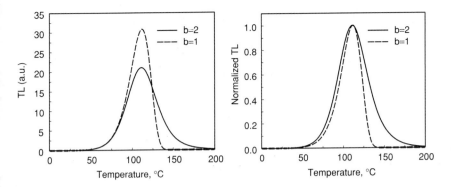

FIGURE 1.3. Schematic comparison of TL glow peaks for first- and second-order kinetics. The parameters are $E = 1$ eV, $s = 10^{12}$ s^{-1}, $n_0 = $ N $= 10^3$ m^{-3}.

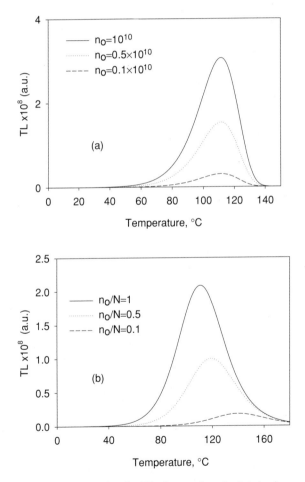

FIGURE 1.4. First-order and second-order TL glow peaks calculated using equations (1.5) and (1.6), and for different initial concentrations n_0 of trapped electrons. The parameters are $E = 1$ eV, $s = 10^{12}$ s^{-1}, N $= 10^{10}$ m^{-3} and (a) $n_0 = 1, 0.5, 0.1 \times 10^{10}$ m^{-3} (b) $n_0/$N $= 1$, 0.5, 0.1.

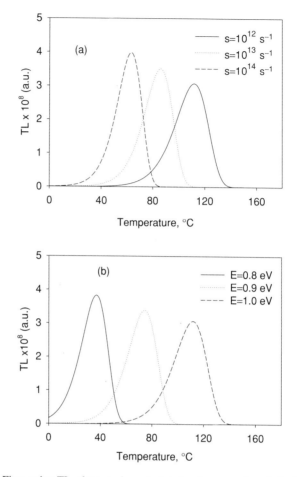

FIGURE 1.5. First-order TL glow peaks calculated using equation (1.5) for (a) different frequency factors s and $E = 1.0$ eV, and (b) for different activation energies E and $s = 10^{12}$ s^{-1}.

concentration n_0, while T_M for second-order TL glow curves shifts toward higher temperatures as n_0 decreases.

Figure 1.5 shows several first-order TL glow peaks calculated using equation (1.5) for (a) different frequency factors s and (b) different activation energies E. As the energy E is increased or as the value of s is decreased, the TL glow curve shifts toward higher temperatures.

Figure 1.6 shows several general-order TL glow peaks calculated using equation (1.7), for different values of the kinetic order parameter $b = 1.5 - 1.7$. The parameters used are $E = 1$ eV, $s = 10^{12}$ s^{-1}, and $n_0 = N = 1$.

The equation giving the maximum of a glow peak is evaluated by setting the derivative of expressions (1.5)–(1.7) equal to zero, to yield the following

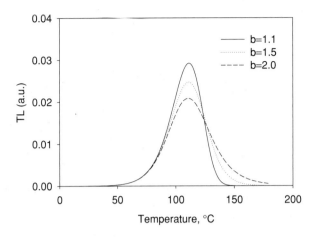

FIGURE 1.6. General order TL glow peaks calculated using equation (1.7) for different values of the kinetic order parameter b. The parameters used are $E = 1$ eV, $s = 10^{12}$ s^{-1}, $\beta = 1$ K s^{-1}, and $n_0 = N = 1$.

equations:

$$\frac{\beta E}{kT_M^2} = s \exp\left(-\frac{E}{kT_M}\right) \qquad \text{first order} \qquad (1.8)$$

$$\frac{\beta E}{kT_M^2} = s \exp\left(-\frac{E}{kT_M}\right)\left[1 + \left(\frac{2kT_M}{E}\right)\right] \qquad \text{second order} \qquad (1.9)$$

$$\frac{\beta E}{kT_M^2} = s \exp\left(-\frac{E}{kT_M}\right)\left[1 + (b-1)\left(\frac{2kT_M}{E}\right)\right] \qquad \text{general order} \qquad (1.10)$$

where T_M is the temperature corresponding to the maximum TL intensity I_M.

It must be noted that the integrals appearing in equations (1.5)–(1.7) can not be calculated in terms of elementary functions, and they must be evaluated using numerical integration methods. Alternatively, an approximating procedure using a series expansion is usually employed, and will be discussed in a later section of this chapter.

The analysis of TL peaks can yield at the most three parameters. The individual TL peaks can be analyzed by using several techniques in order to determine the kinetic TL parameters, such as the activation energy E, the frequency factor s and the order of kinetics b. In an experimental situation the quantities of interest are the temperature T_M where the maximum TL intensity I_M occurs, and the "width" of the TL glow peak. In addition, the knowledge of the inflections points of the TL glow curve can be useful. These inflection points can be computed by setting the second derivative of the data equal to zero. These geometrical properties of the TL glow curves are discussed in a later section of this chapter.

Methods of Analysis

Initial Rise Methods

The experimental methods described in this section apply to any order of kinetics, and are based on the analysis of the low temperature interval of a peak. The initial rise method of analysis was first suggested by Garlick and Gibson [2].

The amount of trapped electrons in the low temperature tail of a TL glow peak can be assumed to be approximately constant, since the dependence of $n(T)$ on temperature T is negligible in that temperature region. In fact, as the temperature increases, the first exponential in equation (1.5) increases, whereas the value of the second term remains close to unity. This remains true for temperatures up to a cutoff temperature T_C, corresponding to a TL intensity I_C smaller than about 15% of the maximum TL intensity I_M. A further increase in temperature ($T > T_C$) makes the second term in equation (1.5) decrease; the competition between the two terms in equation (1.5) results in the peak-shape of the TL glow curve.

By using this assumption of constant $n(T)$, the thermoluminescence emission can be described by

$$I(T) \propto \exp\left(-\frac{E}{kT}\right) \tag{1.11}$$

Figure 1.7 shows the initial rise part of a single TL glow peak. In applying the initial rise method, a graph of $\ln(I)$ versus $1/kT$ is made, and a straight line is obtained. From the slope $-E$ of the line, the activation energy E is evaluated without any knowledge of the frequency factor s. An example of the initial rise plot is given in Figure 1.8.

An alternative method is the graphical method proposed by Ilich [4], which is shown in Figure 1.9. One uses a point I_C on the isolated TL glow peak, draws the tangent at the point $N = (T_C, I_C)$ and calculates the slope, assuming that $I(T)$ is

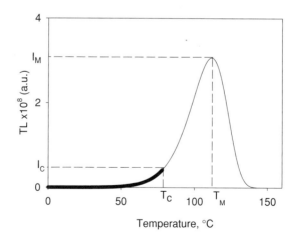

FIGURE 1.7. The initial rise part of a thermoluminescence glow curve.

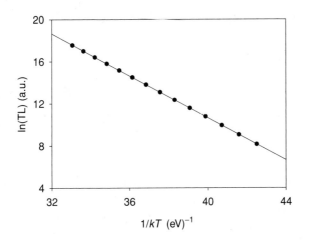

FIGURE 1.8. Applying the initial rise technique to the initial rise portion of Figure 1.7.

given by

$$I(T) = c \exp\left(-\frac{E}{kT}\right) \tag{1.12}$$

The derivative of this expression is equal to

$$\frac{dI}{dT} = c\frac{E}{kT^2}\exp\left(-\frac{E}{kT}\right) = I\frac{E}{kT^2} \tag{1.13}$$

The slope of the tangent at point N in Figure 1.9 is found by setting $T = T_C$ in equation (1.13) to obtain

$$\left.\frac{dI}{dT}\right|_{T=T_C} = I_C\frac{E}{kT_C^2}. \tag{1.14}$$

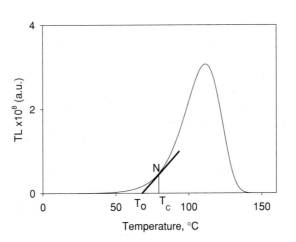

FIGURE 1.9. The graphical method proposed by Ilich.

In order to find the point $M \equiv (T_0, 0)$ where the graph intersects the x-axis, one uses the equation of the tangent line

$$I - I_C = \frac{I_C E}{kT_C^2}(T - T_C)$$
$$I = 0. \tag{1.15}$$

The solution of this system gives the value of E:

$$E = k\frac{T_C^2}{T_C - T_0}. \tag{1.16}$$

A large number of points (T_C, I_C) can be selected and statistically processed in order to improve the reliability of the results of this method.

Aramu et al [5] applied the initial rise method in the case where the frequency factor s depends on the temperature T. In this case the TL intensity is given by

$$I \propto T^\alpha \exp\left(-\frac{E}{kT}\right) \tag{1.17}$$

where $\alpha = $ constant.

From this equation one obtains

$$\frac{d\ln(I)}{dT} = \frac{\alpha}{T} + \frac{E}{kT^2} \tag{1.18}$$

This equation can be compared to the equation obtained in the usual case where the frequency factor s is independent of the temperature T:

$$\frac{d\ln(I)}{dT} = \frac{E_{IR}}{kT^2} \tag{1.19}$$

where E_{IR} is the activation energy obtained using the initial rise method.

By comparing equations (1.18) and (1.19) one obtains the actual activation energy E:

$$E = E_{IR} - \alpha kT. \tag{1.20}$$

Equation (1.20) provides the correct value of E within a few percent, for the case where the frequency factor s depends on the temperature T.

It is pointed out that if the resolution of the TL glow peak is poor due to the presence of nearby overlapping peaks, the initial rise method is not applicable. Such a situation is depicted in Figure 1.10, which shows two different glow peaks, with the second one containing a shoulder on its ascending part. In cases such as the one shown in Figure 1.10, one must attempt to separate the composite TL glow curve into its constituent components before attempting to apply the initial rise method.

Several methods have been suggested in the literature for applying the initial rise method to composite TL glow curves.

The first method involves a thermal cleaning technique [6]. The sample is heated beyond the maximum of the first TL peak, is subsequently cooled to a temperature

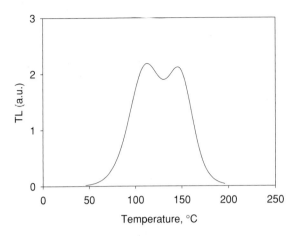

FIGURE 1.10. Example of overlapping TL glow peaks in which the initial rise method may fail.

value where the luminescence is negligible, and subsequently reheated beyond the maximum of the next peak and cooled again. The procedure is repeated for all the peaks. This method should produce clean initial rise curves for each peak; in practice, a complete thermal cleaning of the TL peaks is not assured and the E values may not be very accurate.

The thermal cleaning technique and initial rise method of analysis can be applied in a more systematic manner by using many heating and cooling cycles, each time to a slightly higher temperature T_{stop}, to yield a series of $I(T)$ graphs to be analyzed using the initial rise method. By graphing the activation energies E obtained by this process as a function of T_{stop}, one obtains usually a "staircase" type of graph, termed the "$E - T_{stop}$ graph". The method can be best applied when the TL peaks are sufficiently separated in temperature.

An alternative method was introduced by McKeever [7]: an irradiated sample is heated at a linear rate up to a temperature T_{stop} corresponding to a point on the low temperature tail of the first peak. The sample is then cooled and reheated to obtain all of the glow curve, and the temperature of maximum TL intensity (T_M) is noted. The procedure is repeated several times by re-irradiating the same sample, or by using a different irradiated sample, using each time a slightly higher value of T_{stop} (each time T_{stop} is increased by 2 to 5°C). A plot of T_M versus T_{stop} shows a stepwise curve with each "flat" region corresponding to a different activation energy E.

The method is applicable to single or overlapping first- and second-order peaks, in which case a smooth staircase structure is produced. When closely overlapping peaks or a quasi-continuous distribution of TL peaks is present, a smooth line of slope about 1 is produced for both first- and second-order cases. This method allows the estimation of the number and position of individual peaks within a complex glow curve. The trapping parameters can then be estimated by applying a computerized curve-fitting procedure to the glow curve.

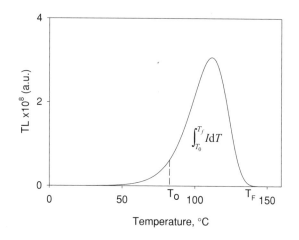

FIGURE 1.11. Calculation of the area $n(T)$ in the whole glow peak—area measurement method.

A more accurate but also experimentally very demanding technique, called the fractional glow (FG) method, has been introduced by Gobrecht and Hofmann [8]. In this case the heating and cooling are done in small temperature intervals. The activation energy E is calculated for each heating and cooling cycle from the $\ln(I)$ versus $1/kT$ plot.

It is noted that the initial rise method of analysis may be affected by the presence of the phenomenon of thermal quenching, which is discussed in Chapter 5 of this book.

Methods of Analysis Employing the Whole TL Glow Curve

These methods are known as "area methods" or "whole glow peak" methods of analysis, and are based on the measurement of the integral under a glow peak; they can be applied when a well-isolated and clean peak is available.

The value of the integral $n(T)$ of the TL intensity over a certain temperature region can be estimated by the area under the glow curve from a given temperature T_0 in the initial rise region, up to the final temperature T_f at the end of the glow peak, as shown in Figure 1.11.

$$n = \int_{t_0}^{t_f} I \, \mathrm{d}t = \frac{1}{\beta} \int_{T_0}^{T_f} I \, \mathrm{d}T. \tag{1.21}$$

Assuming first-order kinetics, and by substituting the Randall–Wilkins relations (1.2) leads to

$$\ln \left[\frac{I}{\int_T^{T_f} I \, \mathrm{d}T} \right] = \ln \frac{s}{\beta} - \frac{E}{kT}. \tag{1.22}$$

This equation shows that in the case of first-order kinetics the term $\ln(I/n(T))$ is a linear function of $1/kT$, with a slope $-E$ and an intercept equal to $\ln(s/\beta)$.

May and Partridge [3] and Muntoni et al [9] proposed the same method in the case of general order kinetics. In this case the equation is

$$\ln\left(\frac{I}{n^b}\right) = \ln\frac{s'}{\beta} - \frac{E}{kT}, \qquad (1.23)$$

which is graphically processed by plotting $\ln(I/n^b)$ versus $1/kT$.

If the kinetic order b is known, one can obtain a broad range of temperatures in which the curve is a straight line. When the kinetic order is unknown, several lines are drawn with various values of b and the best straight line is chosen.

Peak Position Methods of Analysis

These methods fall under two broad categories, estimation methods based on the location of maximum TL intensity T_M, and methods which employ variable heating rates during measurement of the TL glow peaks.

Methods of Analysis Based on the Temperature at the Maximum

Randall and Wilkins [1] did not solve the first-order equation, but they considered the maximum temperature of the TL glow peak as corresponding to a temperature slightly below that in which the probability of an electron escaping from the trap is equal to unity. These authors found a very simple expression for E, using a value of $s = 2.9 \times 10^9$ s^{-1}:

$$E = 25kT_M. \qquad (1.24)$$

Urbach [10] gave a similar relation using $s = 10^9$ s^{-1}:

$$E = \frac{T_M}{500} = 23kT_M. \qquad (1.25)$$

The numerical factors in both equations (1.24) and (1.25) are dependent upon the s value, and hence the values of E are only approximate. These equations can be used only as a first approximation of the E-values.

Methods of Analysis Based on Various Heating Rates

When the linear heating rate β changes, the temperature T_M of the maximum TL intensity of the peak also changes: faster heating rates produce a shift in temperature toward higher values of T_M. This effect is shown in Figure 1.12.

Bohum [11], Porfianovitch [12] and Booth [13] proposed a method of calculating E based on two different heating rates for a first-order peak. Considering the

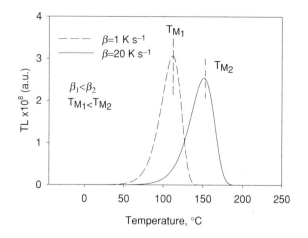

FIGURE 1.12. The change in the temperature T_M of maximum intensity with the heating rate.

maximum condition equation (1.8) and using two different heating rates, one obtains

$$E = k \frac{T_{M1} T_{M2}}{T_{M1} - T_{M2}} \ln \left[\frac{\beta_1}{\beta_2} \left(\frac{T_{M2}}{T_{M1}} \right)^2 \right] \tag{1.26}$$

If T_M can be measured within an accuracy of $1°C$, this method yields E within an accuracy of 5%.

In a slightly different method, Hoogenstraaten [14], starting from the first-order equation, suggested the use of several heating rates to obtain the following linear relation:

$$\ln \left(\frac{T_M^2}{\beta} \right) = \frac{E}{kT_M} + \ln \left(\frac{E}{sk} \right). \tag{1.27}$$

The resultant plot of $\ln(T_M^2/\beta)$ versus $1/kT_M$ should yield a straight line with slope E and an intercept $\ln(E/sk)$.

Chen and Winer [15] reported a method which uses an approximation for the integral appearing in the general-order expression of $I(T)$, obtaining the following equation:

$$\ln \left[I_M^{b-1} \left(\frac{T_M^2}{\beta} \right)^b \right] = \frac{E}{kT_M} + c \tag{1.28}$$

where $c =$ constant.

By means of this equation it is possible to evaluate the quantity on the left side for different values of b, and to obtain a set of graphs as a function of $1/kT_M$. The value of b for which the graph best approximates linearity is found, and the graphs are fitted by a straight line whose slope is E. The method is valid for general heating rates, i.e. the heating rate β does not need to be constant.

For the case of second-order kinetics the above equation becomes

$$\ln\left[I_M\left(\frac{T_M^2}{\beta}\right)^2\right] = \frac{E}{kT_M} + c. \tag{1.29}$$

This method is useful only when b is appreciably different from unity, since for $b = 1$ the temperature T_M of maximum TL intensity is independent of the initial concentration n_0 of trapped electrons.

Chen and Winer [15] used the condition of maximum emission and the integral approximation, and they obtained

$$\left(\frac{\beta}{T_M^2}\right) \cong \exp\left(-\frac{E}{kT_M}\right)\left(\frac{ks}{E}\right)[1 + (b-1)\Delta_M] \tag{1.30}$$

where $\Delta_M = 2kT_M/E$. The quantity $[1 + (b-1)\Delta_M]$ is close to unity and can be considered a constant, so that the plot of $\ln(\beta/T_M^2)$ versus $1/kT_M$ should yield a straight line of slope $-E$.

A different method that uses two heating rates was proposed in [16]. It is analogous to the Booth method, which is strictly valid for a first-order peak, but in this case it is applied to a non-first-order TL peak and it is based on the variation of I_M with the heating rate β, which is much faster than the variation of T_M with β. Using the general-order expression one obtains

$$E = \frac{kT_{m1}T_{m2}}{T_{m1} - T_{m2}}\ln\frac{I_{m1}}{I_{m2}}. \tag{1.31}$$

The maximum systematic error in the activation energy E when using equation (1.31) is less than 1% for any order of kinetics ($1.1 \le b \le 2.5$).

Chen and Winer [15] showed that in the case of a temperature-dependent pre-exponential factor s (s proportional to T^α), the graph of $\ln(T_M^2/\beta)$ versus $1/kT_M$ should yield a straight line of slope $E + \alpha kT_M$ instead of the actual activation energy E.

It must be emphasized that during application of the variable heating rate methods of analysis, it is essential to have good thermal contact between the heating element in the TL apparatus and the sample. An example of correcting experimental data for temperature lag effects is given in Chapter 5.

Isothermal Decay Method

In this section a method of analysis that does not employ any particular heating cycle will be discussed. The usual experimental procedure in an isothermal decay experiment consists of quickly heating the sample after irradiation to a specific temperature, and keeping the sample at this temperature for a given time interval. The light output (also termed phosphorescence decay) is measured as a function of time, and so it is possible to evaluate the decay rate of trapped electrons. Graphs of the TL intensity versus time t at constant temperature are called *isothermal decay curves*.

The method of isothermal decay analysis was illustrated for first-order kinetics by Garlick and Gibson [2]. The isothermal decay curves at a temperature T_i for TL peaks following first-order kinetics are exponential graphs as a function of time, given by

$$ I = I_0 \exp \left(-s \exp \left(-\frac{E}{kT_i} \right) t \right). \tag{1.32} $$

This equation indicates that a graph of $\ln(I)$ versus time will be linear for first-order kinetics peaks, and that the slope of the line will be

$$ \text{slope} = m_i = -s \exp \left(-\frac{E}{kT_i} \right). \tag{1.33} $$

By taking the natural logarithm of this equation we obtain

$$ \ln(|\text{slope}|) = \ln s - \frac{E}{kT_i}. \tag{1.34} $$

The graph of the $\ln(|\text{slope}|)$ versus $1/kT$ should be a straight line with slope $= -E$ and a Y-intercept equal to $\ln s$.

If the experiment is carried out with two different constant temperatures, T_1 and T_2, two different slopes m_1 and m_2 are obtained and equation (1.34) gives

$$ \ln \left(\frac{m_1}{m_2} \right) = \frac{E}{k} \left(\frac{1}{T_2} - \frac{1}{T_1} \right). \tag{1.35} $$

This equation can be used to calculate E.

The application of isothermal decay analysis for general-order kinetics has been suggested in [3], [17]. By using isothermal analysis in this case it is also possible to find the order of kinetics b. By keeping the temperature constant and by integrating the general-order equation (1.4) with respect to time t, one gets

$$ I_t = I_0 \left[1 + s'n_0^{b-1}(b-1)t \exp \left(-\frac{E}{kT} \right) \right]^{\frac{b}{1-b}} \tag{1.36} $$

where

$$ I_0 = s'n_0^b \exp \left(-\frac{E}{kT} \right). \tag{1.37} $$

I_0 and n_0 are, respectively, the initial TL intensity and the initial concentration of trapped charges and I_t is the TL intensity at time t. By rearranging equation (1.36) we obtain

$$ \left(\frac{I_t}{I_0} \right)^{\frac{1-b}{b}} = \left[1 + s'n_0^{b-1}(b-1)t \exp \left(-\frac{E}{kT} \right) \right] \tag{1.38} $$

This equation indicates that a plot of the quantity $(I_t/I_0)^{\frac{1-b}{b}}$ versus time should be a straight line when a suitable value of b is found.

Using different isothermal decay temperatures, a set of straight lines of slopes

$$ m = s'n_0^{b-1}(b-1)\exp \left(-\frac{E}{kT} \right) \tag{1.39} $$

is obtained and the activation energy E can be determined from the plot of $\ln(m)$ versus $1/kT$.

Alternatively, b can be determined using the expression of May and Partridge [3]:

$$\ln\left(\frac{dI}{dt}\right) = \ln C + \frac{2b-1}{b}\ln(I). \qquad (1.40)$$

The plot of $\ln(dI/dt)$ versus $\ln(I)$ gives a straight line having a slope $m = (2b-1)/b$ from which b can be evaluated.

Thermoluminescence materials may exhibit isothermal decay behaviors which do not follow the expressions for first-, second- and general-order kinetics. They may follow instead a decay law of the form $t^{-\alpha}$. For example, a commonly observed decay law is a $1/t$-law, which has been attributed to a uniform distribution of energies [18]. Another well-known example of such a law, which is temperature independent, has been observed in calcite and has been interpreted as due to a quantum mechanical tunneling effect [19].

Methods of Analysis Based on the Shape of the Glow Curve

A popular method of analyzing a TL glow curve in order to ascertain the kinetic parameters E, s, and b is by considering the shape or geometrical properties of the peak. TL glow peaks corresponding to second-order kinetics are characterised by an almost symmetrical shape, whereas first-order peaks are asymmetrical. One defines the following parameters shown in Figure 1.13:

- T_M is the peak temperature at the maximum
- T_1 and T_2 are, respectively, the temperatures on either side of T_M, corresponding to half intensity
- $\tau = T_M - T_1$ is the half-width at the low temperature side of the peak
- $\delta = T_2 - T_M$ is the half-width toward the fall-off side of the glow peak
- $\omega = T_2 - T_1$ is the total half-width
- $\mu = \delta/\omega$ is the so-called geometrical shape or symmetry factor.

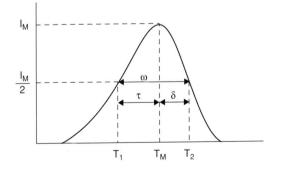

FIGURE 1.13. The geometrical shape quantities τ, δ, ω.

Grossweiner was the first to use the shape of the glow peak to calculate the trap depth E [20]. His method is based on the temperature at maximum intensity T_M and on the low temperature at half intensity, T_1. By assuming first-order kinetics he obtained

$$E = 1.51k\frac{T_M T_1}{T_M - T_1}. \tag{1.41}$$

This expression was empirically modified by Chen [21] with a factor of 1.41 replacing Grossweiner's factor of 1.51, in order to obtain a better accuracy in the calculation of E.

Lushchik [22] also proposed a method based on the shape of the TL glow peak for both first- and second-order kinetics. Introducing the parameter δ defined above, a glow peak can be approximated by a triangle.

For the case of first-order kinetics, the expression for E is

$$E = \frac{kT_M^2}{\delta}. \tag{1.42}$$

The Lushchik formula for second-order kinetics is [20]

$$E = \frac{2kT_M^2}{\delta}. \tag{1.43}$$

Chen [21] modified the two previous equations in order to get a better accuracy in the E value, by multiplying equation (1.42) by 0.978, and equation (1.43) by 0.853.

Halperin and Braner [23] gave a different approach by using both T_1 and T_2 on the glow curve:

$$E = \frac{1.72}{\tau}kT_M^2 (1 - 2.58\Delta_M) \quad \text{for first order} \tag{1.44}$$

$$E = \frac{2}{\tau}kT_M^2 (1 - 3\Delta_M) \quad \text{for second order} \tag{1.45}$$

with
$$\Delta_M = \frac{2kT_M}{E}.$$

The equations of Halperin and Braner require an iterative process in order to find E, due to the presence of Δ_M which also depends on E. To overcome this difficulty, a new approximating method was proposed by Chen [21] who obtained the expressions

$$E = 2kT_M \left(1.25\frac{T_M}{\omega} - 1 \right) \quad \text{for first order.} \tag{1.46}$$

$$E = 2kT_M \left(1.76\frac{T_M}{\omega} - 1 \right) \quad \text{for second order.} \tag{1.47}$$

Chen [24] also derived general expressions for evaluating E. His method is useful for a broad range of energies ranging between 0.1 eV and 2.0 eV, and for values of the pre-exponential factors between 10^5 s^{-1} and 10^{23} s^{-1}. Furthermore, Chen's

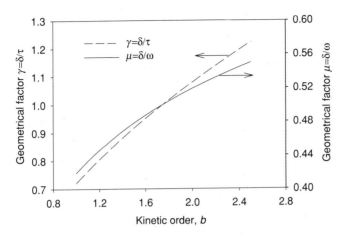

FIGURE 1.14. Relationship between the kinetic order b and the geometrical factors $\mu = \delta/\omega$ and $\gamma = \delta/\tau$.

method does not make use of any iterative procedures and does not require knowledge of the kinetic order, which is found by using the symmetry factor μ from the peak shape. The equations can be summed up as

$$E_\alpha = c_\alpha \left(\frac{kT_M^2}{\alpha} \right) - b_\alpha (2kT_M) \qquad (1.48)$$

where α is τ, δ, or ω. The values of c_α and b_α are summarized as below:

$$c_\tau = 1.510 + 3.0(\mu - 0.42) \qquad b_\tau = 1.58 + 4.2(\mu - 0.42)$$
$$c_\delta = 0.976 + 7.3(\mu - 0.42) \qquad b_\delta = 0$$
$$c_\omega = 2.52 + 10.2(\mu - 0.42) \qquad b_\omega = 1, \qquad (1.49)$$

with $\mu = 0.42$ for the case of first-order TL glow peaks, and $\mu = 0.52$ for the case of second-order peaks.

Chen [24] calculated a graph of μ ranging from 0.36 to 0.55, for values of b between 0.7 and 2.5, which can be used for the evaluation of b from a measured μ as shown in Figure 1.14. Another graph has been proposed by Balarin [25] which gives the kinetics order as a function of the parameter $\gamma = \delta/\tau$, and which is also shown in Figure 1.14.

Curve Fitting Methods

The Series Approximation to the TL Integrals

The integral appearing in expression (1.5) for first-order TL glow peaks is

$$F(T, E) = \int_{T_0}^{T} \exp(-E/kT') dT'. \qquad (1.50)$$

This integral cannot be solved analytically, but can be found by successive integration by parts to be equal to [26]

$$F(T, E) = T \exp(-E/kT) \sum_{n=1}^{\infty} \left(\frac{kT}{E}\right)^n (-1)^n n!$$ (1.51)

By keeping only the first two terms in this approximation, the integral can be approximated by

$$F(T, E) = \frac{kT^2}{E} \exp(-E/kT) \left(\frac{1 - 2kT}{E}\right)$$ (1.52)

By using this approximation to the integral and by further using the condition for the temperature T_M of the maximum TL intensity equation (1.8), the following expression can be derived for the TL intensity of first-order TL glow curves [26]:

$$I(T) = I_M \exp\left[1 + \frac{E}{kT}\frac{T - T_M}{T_M} - \frac{T^2}{T_M^2}\left(1 - \frac{2kT_M}{E}\right)\right.$$
$$\left. \times \exp\left(\frac{E}{kT}\frac{T - T_M}{T_M}\right) - \frac{2kT_M}{E}\right]$$ (1.53)

This equation will be referred to in the rest of this book as the *Kitis et al equation* for first-order kinetics, and will be used in Chapter 2 to analyze the first-order TL glow curves. The advantage of using this equation to approximate equation (1.5) is that it involves two quantities which are measured experimentally: T_M and I_M. The activation energy E can be treated as an adjustable parameter.

By using the series approximation in equations (1.6) and (1.7), the following expressions can be derived in a similar manner for second- and general-order TL glow curves [26]:

$$I(T) = 4I_M \exp\left(\frac{E}{kT}\frac{T - T_M}{T_M}\right)$$
$$\times \left[\frac{T^2}{T_M^2}\left(1 - \frac{2kT}{E}\right)\exp\left(\frac{E}{kT}\frac{T - T_M}{T_M}\right) + 1 + \frac{2kT_M}{E}\right]^{-2}$$ (1.54)

$$I(T) = I_M b^{\frac{b}{b-1}} \exp\left(\frac{E}{kT}\frac{T - T_M}{T_M}\right)\left[1 + (b - 1)\frac{2kT_M}{E} + (b - 1)\right.$$
$$\left. \times \left(1 - \frac{2kT}{E}\right)\left(\frac{T^2}{T_M^2}\exp\left(\frac{E}{kT}\frac{T - T_M}{T_M}\right)\right)\right]^{\frac{-b}{b-1}}$$ (1.55)

These expressions will be used in Chapter 2 to evaluate the activation energy E for general-order kinetics by using a curve fitting procedure.

Computerized Curve Fitting Procedures

The subject of computerized curve fitting analysis became very popular during the last decade with the development of sophisticated glow curve deconvolution techniques (GCD). Some simple examples of curve fitting methods that can be applied to single TL glow peaks are given in Chapters 2 and 3.

A detailed presentation of this important subject is beyond the scope of this book, and for detailed reviews the reader is referred to the annotated bibliography at the end of this book. Only a few general relevant comments will be presented in this section.

Chen and McKeever [19] have summarized the curve fitting procedures commonly used to analyze multipeak TL glow curves. They emphasize the primary importance of using a carefully measured TL glow curve, since any errors in measuring the glow curve can lead to the wrong results in the computerized procedures. Such procedures are more likely to yield accurate results in the case of linear superposition of first-order Randall–Wilkins-type mathematical expressions. These authors concluded that curve fitting methods using a particular theoretical model should be applied with the utmost care, and extreme caution should be exercised when drawing conclusions from good curve fitting results.

References

[1]   J.T. Randall and M.H.F. Wilkins, *Proc. Roy. Soc.* A **184** (1945) 366

[2]   G.F.J. Garlick and A.F. Gibson, *Proc. Phys. Soc.* **60** (1948) 574

[3]   C.E. May and J.A. Partridge, *J. Chem. Soc.* **40** (1964) 1401

[4]   B.M. Ilich, *Sov. Phys. Solid State*, **21** (1979) 1880.

[5]   F. Aramu, P. Brovetto, and A. Rucci, *Phys Lett.* **23** (1966) 308.

[6]   K.H. Nicholas and J. Woods, *Br. J. Appl. Phys.* **15** (1964) 783

[7]   S.W.S. McKeever, *Phys. Status Solidi* a, **62** (1980) 331.

[8]   H. Gobrecht and D. Hofmann, *J. Phys. Chem. Solids* **27** (1966) 509

[9]   C. Muntoni, A. Rucci, and A. Serpi, *Ricerca Scient.* **38** (1968) 762

[10] F. Urbach, *Winer Ber. IIa* **139** (1930) 363.

[11] A. Bohum, *Czech. J. Phys.* **4** (1954) 91.

[12] I.A. Porfianovitch, *J. Exp. Theor. Phys. USSR* **26** (1954) 696.

[13] A.H. Booth, *Canad. J. Chem.* **32** (1954) 214.

[14] W. Hoogenstraaten, *Philips Res. Rep.* **13** (1958) 515.

[15] R. Chen and S.A.A. Winner, *J. Appl. Phys.* **41** (1970) 5227.

[16] R.K. Gartia, S. Ingotombi, T.S.G. Singh, and P.S. Mazmudar, *J. Phys. D: Appl. Phys.* **24** (1991) 65

[17] N. Takeuchi, K. Inabe, and H. Nanto, *J. Mater. Sci.* **10** (1975) 159.

[18] S.W.S. McKeever 1988 *Thermoluminescence of Solids.* Cambridge: Cambridge University Press.

[19] R. Chen and S.W.S. McKeever 1997 *Theory of Thermoluminescence and Related Phenomena.* Singapore: World Scientific.

[20] L.I. Grossweiner, *J. Appl. Phys.* **24** (1953) 1306.

[21]   R. Chen, *J. Appl. Phys.* **40** (1969) 570.
[22]   L.I. Lushihik, *Soviet Phys. JEPT* **3** (1956) 390.
[23]   A. Halperin and A.A. Braner, *Phys. Rev.* **117** (1960) 408
[24]   R. Chen, *J. Electrochem. Soc.* **116** (1969) 1254.
[25]   M. Balarin, *Phys. Status Solidi* a, **54** (1979) K137.
[26]   G. Kitis, J.M. Gomez-Ros, and J.W.N. Tuyn, *J. Phys.* D **31** (1998) 2636.

2
Analysis of Thermoluminescence Data

Introduction

In this chapter, the analytical expressions presented in Chapter 1 will be used in several detailed numerical exercises. A variety of methods will be used to analyze the same TL glow-curve data, and the results from the different methods will be compared with each other.

Chen and McKeever [1] have provided an excellent summary of how to systematically analyze TL glow curves, by following these steps:

(1) Ensure that the temperature measurement during the TL glow peak is accurate, by optimizing the thermal contact between the sample and the heating element.

(2) Eliminate the possibility of nearby overlapping peaks, by using a thermal cleaning process. Thermal quenching effects must also be accounted for, if present, and corrected using theoretical considerations. The study of emission spectra during the TL glow curve provides also valuable information about the TL process.

(3) Characterize the isolated glow peak by evaluating the three parameters E, s, and b using several of the standard methods of analysis. Methods utilizing the whole glow peak should be preferred over methods based on only a few points on the glow curve. It is essential to carry out this analysis for different trap filling, by studying, for example, samples irradiated at different doses.

(4) It is important to get good agreement between several methods of analysis. Any discrepancies should be examined in more detail.

(5) In order to resolve discrepancies and obtain more information about the processes involved, the analysis should be carried out for glow peaks measured under different heating rates, various irradiation doses, powdered and bulk samples, etc.

(6) Additional information should be obtained using experimental methods based on different physical processes, such as isothermal decay techniques, dose-dependence measurements, excitation and emission spectra, and simultaneous TL-TSC (thermally stimulated current) measurements.

Exercises 2.1–2.3 contain a detailed example of analyzing first-order TL data using three different methods: peak shape methods, variable heating rate methods, and isothermal decay techniques. Similarly, examples of analyzing second- and general-order TL glow curves are given in Exercises 2.4–2.6. Although the material in these exercises may seem to be repeated at times, we have chosen to provide complete and self-contained exercises for easy reference, instead of constantly referring the reader to previous sections of the book.

Extra attention has been paid to include an error analysis of the data whenever possible, because there seems to be a general lack of such detailed examples of error analysis in the TL literature. Exercises 2.7 and 2.9 present examples of the effect of experimental background on the accuracy of the initial rise (IR) method, and of the propagation of errors in the peak shape methods of analysis. Exercise 2.8 is a simulated study of the well-known "15% TL intensity" rule of thumb which is commonly used in experimental TL work.

Exercise 2.1: Analysis of a First-Order TL Peak

You are given the experimental data in Table 2.1 and Figure 2.1, for a TL glow curve (TL versus temperature T), and a known heating rate $\beta = 1$ K s^{-1}.

(a) Apply the IR method to find the activation energy E. The value for E obtained using the IR method is assumed to be independent of the order of kinetics.

(b) Apply Chen's peak shape equations to find the activation energy E, using the shape parameters τ, δ, and ω. By assuming that the experimental error in the quantities τ, δ, and ω is $\Delta T = 2$ K, estimate the error $\Delta \mu$ in the value of the geometrical shape factor μ.

Show that the values of μ and $\Delta \mu$ are consistent with the assumption that the TL glow curve obeys first-order kinetics.

(c) By using the experimental data, apply the whole glow-peak method to find E, s, and the order of kinetics b. Graph $\ln(I/n^b)$ versus $1/T$ for various values of

TABLE 2.1. The experimental data for a first-order TL glow curve

T(C)	TL$_{\text{experimental}}$	T(C)	TL$_{\text{experimental}}$
20	7.56×10^4	85	8.07×10^7
25	1.28×10^5	90	1.20×10^8
30	2.44×10^5	95	1.71×10^8
35	4.54×10^5	100	2.31×10^8
40	8.29×10^5	105	2.90×10^8
45	1.49×10^6	110	3.26×10^8
50	2.61×10^6	115	3.15×10^8
55	4.51×10^6	120	2.43×10^8
60	7.65×10^6	125	1.34×10^8
65	1.27×10^7	130	4.49×10^7
70	2.09×10^7	135	6.75×10^6
75	3.35×10^7	140	2.57×10^5
80	5.27×10^7	145	2.73×10^3

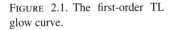

FIGURE 2.1. The first-order TL glow curve.

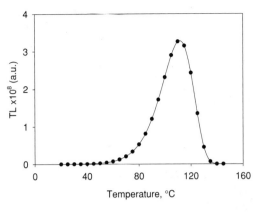

b between 0.8 and 1.2, and find the correct value of b that gives a linear graph. From the slope and intercept of the graph $\ln(I/n^b)$ versus $1/T$ calculate E and s. Verify that the given TL glow curve corresponds to first-order kinetics.

(d) Using the experimental values of I_M (maximum TL intensity) and T_M (temperature of maximum intensity), do a curve fitting to the given numerical data. Use the following analytical equation developed by Kitis et al for first-order kinetic peaks [2]. The expression relies on two experimentally measured quantities, I_M and T_M:

$$I(T) = I_M \exp\left[1 + \frac{E}{kT} \cdot \frac{T - T_M}{T_M} - \frac{T^2}{T_M^2}\right.$$
$$\left. \times \left(1 - \frac{2kT_M}{E}\right) \exp\left(\frac{E}{kT} \cdot \frac{T - T_M}{T_M}\right) - \frac{2kT_M}{E}\right]. \quad (2.1)$$

The activation parameter E can be treated as an adjustable parameter in this expression. Graph both the experimental data and the calculated first-order TL glow curve on the same graph and compare them. Calculate the figure of merit (FOM) of the TL glow curve.

(e) Can it be concluded for this material that this TL peak will always follow first-order kinetics?

Solution

(a) *The IR method.* We calculate in Table 2.2. the values of $1/kT$ (T = temperature in K, k = Boltzman constant) and the values of the natural logarithm of the TL data, $\ln(\mathrm{TL})$ in a spreadsheet.

We next graph the $\ln(\mathrm{TL})$ versus $1/kT$ data as shown in Figure 2.2.

A very important consideration when applying the IR method is deciding how many data points to use for the regression analysis of the graph $\ln(\mathrm{TL})$ versus $1/kT$. We obtain the activation energy E by graphing $\ln(\mathrm{TL})$ versus $1/kT$ for the initial part of the data. By performing a regression line analysis using the *first 16 data points* up to a temperature of 100°C, we obtain the graph in Figure 2.3.

TABLE 2.2. Calculated values of $1/kT$ and the values of ln(TL) for first-order glow curve

T(C)	TL$_{\text{experimental}}$	$1/kT(\text{eV}^{-1})$	ln(TL)	T(C)	TL$_{\text{experimental}}$	$1/kT(\text{eV}^{-1})$	ln(TL)
20	7.56×10^4	39.47	11.23	85	8.07×10^7	32.42	18.11
25	1.28×10^5	38.94	11.66	90	1.20×10^8	31.97	18.60
30	2.44×10^5	38.30	12.41	95	1.71×10^8	31.54	18.86
35	4.54×10^5	37.68	12.93	100	2.31×10^8	31.11	19.26
40	8.29×10^5	37.08	13.63	105	2.90×10^8	30.70	19.38
45	1.49×10^6	36.49	14.11	110	3.26×10^8	30.30	19.60
50	2.61×10^6	35.93	14.78	115	3.15×10^8	29.91	19.47
55	4.51×10^6	35.38	15.22	120	2.43×10^8	29.53	19.31
60	7.65×10^6	34.85	15.85	125	1.34×10^8	29.16	18.62
65	1.27×10^7	34.33	16.26	130	4.49×10^7	28.80	17.62
70	2.09×10^7	33.83	16.85	135	6.75×10^6	28.44	15.62
75	3.35×10^7	33.35	17.23	140	2.57×10^5	28.10	12.45
80	5.27×10^7	32.88	17.78	145	2.73×10^3	27.76	7.91

When the first 16 data points are used (intensity up to a temperature of 100°C, corresponding to a TL intensity equal to approximately 70% of the maximum TL intensity), the value of the activation energy $E = 0.976 \pm 0.004$ eV is obtained with a value of the regression coefficient $R^2 = 0.9997$.

By performing a similar regression line analysis using only the *first 11 data points* up to a temperature of 75°C, we obtain the graph in Figure 2.4.

When the first 11 data points are used (up to a temperature of 75°C, corresponding to a TL intensity equal to approximately 9% of the maximum TL intensity), a value of $E = 0.986 + 0.003$ eV is obtained, with an R^2 value of 0.9997. Both graphs in Figures 2.3 and 2.4 give an equally good fit with the same value of R^2.

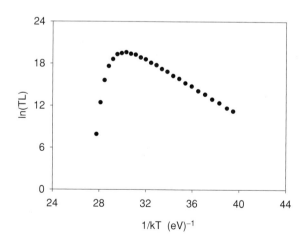

FIGURE 2.2. The IR method applied to the first-order TL glow curve.

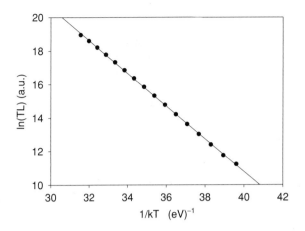

FIGURE 2.3. Applying the IR method to the first 16 experimental points.

The example above shows that the value of the activation energy E obtained from the IR method by doing a regression analysis of the data is very sensitive to the number of points used in the analysis. Exercise 2.7 is an example of the influence of the experimental background on the results of the IR method.

As a general practical rule, application of the IR technique should be restricted to the portion of the TL glow curve corresponding to about 5–10% of the maximum TL intensity. Exercise 2.8 is a detailed simulation of this so-called "15% intensity" rule of thumb commonly used in the IR method.

(b) *Chen's peak shape equations.* From the given experimental data for a TL glow peak, we can estimate the three temperatures required for Chen's peak shape equations:

$$T_1 = 92°C = 365\,K, \quad T_2 = 122°C = 395\,K, \quad T_M = 110°C = 383\,K,$$

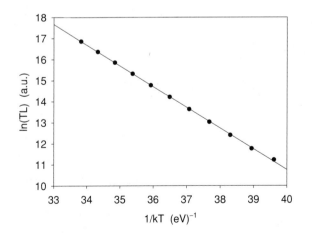

FIGURE 2.4. Applying the IR method to the first 11 experimental points.

where

T_M = peak temperature at the maximum TL intensity,

T_1, T_2 = temperatures on either side of T_M, corresponding to the half-maximum intensity.

We first calculate the quantities μ, τ, δ, and ω:

$$\tau = T_M - T_1 = 18 \, \text{K},$$
$$\delta = T_2 - T_M = 12 \, \text{K},$$
$$\omega = T_2 - T_1 = 30 \, \text{K},$$
$$\mu = \delta/\omega = 12/30 = 0.40.$$

Using the value of τ:

$$E = \frac{1.51kT_M^2}{\tau} - 1.58(2kT_M) = 1.060 - 0.104 = 0.956 \, \text{eV}.$$

Using the value of δ:

$$E = \frac{0.976kT_M^2}{\delta} = 1.028 \, \text{eV}.$$

Using the value of ω:

$$E = \frac{2.52kT_M^2}{\omega} - 2kT_M = 1.062 - 0.066 = 0.996 \, \text{eV}.$$

The value of the geometrical shape factor $\mu = 0.40$ is very close to the value expected for a first-order TL peak which is equal to $\mu = \delta/\omega = 0.42$.

Using the known experimental error $\Delta T = 2$ K or the quantities τ, δ, and ω, we can do an error analysis of the values of μ. By taking the logarithmic derivative of the equation $\mu = \delta/\omega$, we find the relative error $\Delta\mu/\mu$:

$$\ln \mu = \ln \delta - \ln \omega$$

$$\left| \frac{\Delta\mu}{\mu} \right| = \left| \frac{\Delta\delta}{\delta} \right| + \left| \frac{\Delta\omega}{\omega} \right| = \left| \frac{2}{12} \right| + \left| \frac{2}{30} \right| = 0.167 + 0.067 = 0.234.$$

This leads to a value of $\mu \pm \Delta\mu = 0.40 \pm 0.09$. This value is consistent with the assumption of first-order kinetics.

In order to estimate the error ΔE in the activation energy E, we take the logarithmic derivative of the equation $E = 0.976kT_M^2/\delta$:

$$\left| \frac{\Delta E}{E} \right| = 2 \left| \frac{\Delta T_M}{T_M} \right| + \left| \frac{\Delta\delta}{\delta} \right| = 2 \left| \frac{2}{383} \right| + \left| \frac{2}{12} \right| = 0.010 + 0.167 = 0.177.$$

This gives a rather large error in $\Delta E = 0.177E = 0.177(1.028) = 0.18$ eV.

A much more detailed error analysis of the peak shape equations is given in Exercise 2.9.

(c) *The whole glow-peak method.* In the whole glow-peak area method, the area $n(T)$ under the glow peak is calculated starting at temperature T, to the maximum temperature available, as shown in Figure 2.5. In the data shown in Table 2.1, the maximum available temperature is $145°$C.

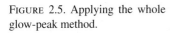

FIGURE 2.5. Applying the whole glow-peak method.

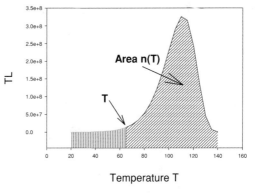

Temperature T

The area $n(T)$ under the glow peak can be approximated by using the sum of all the data points from any temperature T upwards, up to the maximum temperature available in the data. According to equation (1.21), this sum must be multiplied by the temperature interval ΔT and must be also divided by the heating rate β. In our case, we are given $\Delta T = 5$ K and $\beta = 1$ K s^{-1}.

In the spreadsheet example shown in Table 2.3, column C labeled "Area" is calculated using the command:

$$\text{Cell C1} = \text{sum(B1:B26)}^*5/1,$$
$$\text{Cell C2} = \text{sum(B2:B26)}^*5/1, \text{ etc.}$$

Once the "Area" column is calculated, column D labeled "ln(TL/Area)" can be calculated by using the command:

$$\text{Cell D1} = \ln(\text{B1}/\text{C1}),$$
$$\text{Cell D2} = \ln(\text{B2}/\text{C2}), \text{ etc.}$$

Additional columns are created in the spreadsheet for the quantities of $\ln(\text{TL/Area}^b)$ and for several values of the kinetic order $b = 1.2, 1.1, 1.0,$ and 0.9. Not all columns are shown for the sake of saving space.

In Figure 2.6, graphs of $\ln(\text{TL/Area}^b)$ versus $1/kT$ are drawn for several values of the kinetic order $b = 1.2, 1.1, 1.0,$ and 0.9.

The graphs in Figure 2.6 corresponding to $b = 0.9, 1.0,$ and 1.2 clearly deviate from straight lines at low values of $1/kT$, and must be rejected.

The $b = 1.1$ graph has the highest value of R^2 and therefore gives the best fit. The data leads us to conclude that the given TL glow peak is described by $b = 1.1$ kinetics. Due to experimental uncertainties in the data and also due to the fact that only 27 data points are available on the TL glow curve, we can say that to a good approximation this can be considered a *first-order* kinetics TL peak. A regression line is fitted to the best line corresponding to $b = 1.1$, to obtain the best slope and the best intercept, as shown in Figure 2.7:

$$\text{Best intercept} = 24.579 \pm 0.11,$$
$$\text{Best slope } E = 0.979 \pm 0.003 \text{ eV}.$$

TABLE 2.3. The quantities $\ln(I/n)$ and $1/kT$ for first-order glow curve

	A	B	C	D	E	
	T(C)	TL$_{experimental}$	Area	ln(TL/Area)	$1/kT$	ln(TL/Area$^{1.1}$)
1	20	7.56×10^4	1.05×10^{10}	−11.84	39.61	−14.15
2	25	1.28×10^5	1.05×10^{10}	−11.31	38.94	−13.62
3	30	2.44×10^5	1.05×10^{10}	−10.67	38.30	−12.98
4	35	4.54×10^5	1.05×10^{10}	−10.05	37.68	−12.35
5	40	8.29×10^5	1.05×10^{10}	−9.45	37.08	−11.75
6	45	1.49×10^6	1.05×10^{10}	−8.86	36.49	−11.17
7	50	2.61×10^6	1.05×10^{10}	−8.30	35.93	−10.61
8	55	4.51×10^6	1.05×10^{10}	−7.75	35.38	−10.06
9	60	7.65×10^6	1.04×10^{10}	−7.22	34.85	−9.53
10	65	1.27×10^7	1.04×10^{10}	−6.71	34.33	−9.01
11	70	2.09×10^7	1.03×10^{10}	−6.21	33.83	−8.51
12	75	3.35×10^7	1.02×10^{10}	−5.72	33.35	−8.03
13	80	5.27×10^7	1.01×10^{10}	−5.25	32.88	−7.56
14	85	8.07×10^7	9.81×10^9	−4.80	32.42	−7.10
15	90	1.20×10^8	9.41×10^9	−4.36	31.97	−6.66
16	95	1.71×10^8	8.81×10^9	−3.94	31.54	−6.23
17	100	2.31×10^8	7.95×10^9	−3.54	31.11	−5.82
18	105	2.90×10^8	6.80×10^9	−3.16	30.70	−5.42
19	110	3.26×10^8	5.35×10^9	−2.80	30.30	−5.04
20	115	3.15×10^8	3.72×10^9	−2.47	29.91	−4.67
21	120	2.43×10^8	2.15×10^9	−2.18	29.53	−4.33
22	125	1.34×10^8	9.31×10^8	−1.94	29.16	−4.00
23	130	4.49×10^7	2.60×10^8	−1.75	28.80	−3.69
24	135	6.75×10^6	3.50×10^7	−1.65	28.44	−3.38
25	140	2.57×10^5	1.30×10^6	−1.62	28.10	−3.03
26	145	2.73×10^3				

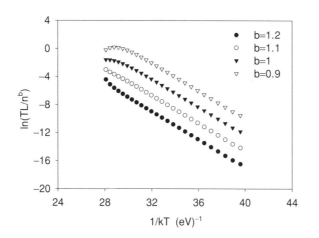

FIGURE 2.6. Applying the whole glow-peak method for different kinetic parameters b.

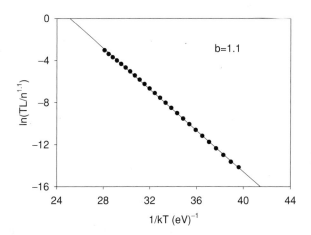

FIGURE 2.7. The value of $b = 1.1$ provides the best least squares fit in the whole glow-curve method of analysis.

This value of E is in reasonably good agreement with the value of $E = 0.986 \pm 0.003$ eV obtained from the IR method. The frequency factor s is calculated from the y-intercept of the graph in Figure 2.7:

$$s = \beta e^{(\text{intercept})} = 1e^{24.579} = 4.7 \times 10^{10} \text{ s}^{-1}.$$

The errors Δs can be calculated from the uncertainties in the intercept of the regression line as follows:

$$\Delta \text{ (intercept)} = \frac{\partial (\ln s)}{\partial s} \Delta s = \frac{\Delta s}{s} = 0.11.$$

This gives a typical large 11% error for the value of the frequency factor, with the final value of s reported as $s = (4.7 \pm 0.5) \times 10^{10} \text{ s}^{-1}$.

(d) *Glow-curve fitting using the Kitis et al equation.* The given TL data can be analyzed by using the following analytical equation developed by Kitis et al [2] for TL peaks following first-order kinetics. The expression relies on two experimentally measured quantities, I_M (the maximum TL intensity) and T_M (the temperature corresponding to the maximum TL intensity):

$$I(T) = I_M \exp \left[1 + \frac{E}{kT} \cdot \frac{T - T_M}{T_M} - \frac{T^2}{T_M^2} \right.$$
$$\left. \times \left(1 - \frac{2kT_M}{E} \right) \exp \left(\frac{E}{kT} \cdot \frac{T - T_M}{T_M} \right) - \frac{2kT_M}{E} \right]. \quad (2.2)$$

For the given experimental data, $T_M = 384$ K and $I_M = 3.26 \times 10^8$. By treating the activation parameter E as an adjustable parameter, we calculate several graphs with values of $E = 0.9, 1.0, 1.1,$ and 1.2 eV. The calculations can be set up easily in a spreadsheet as shown in Table 2.4. Only the first 5 rows are shown for the sake of brevity.

TABLE 2.4. Calculations for glow-curve fitting for several E values

	A	B	C	D	E	F	G	H	I	J	K
			$I(T)$	$I(T)$	$I(T)$	$I(T)$					
1	T(K)	TL$_{\text{experimental}}$	$E = 1$ eV	$E = 0.9$ eV	$E = 1.1$ eV	$E = 1.2$ eV					
2							$E =$	1	eV	0.9	1.1
3	293	7.56×10^4	6.96×10^4	1.77×10^5	2.74×10^4	1.08×10^4					
4	298	1.28×10^5	1.35×10^5	3.21×10^5	5.69×10^4	2.39×10^4					
5	303	2.44×10^5	2.57×10^5	5.72×10^5	1.15×10^5	5.17×10^4	$T_{\text{M}} =$	384	K		
6	308	4.54×10^5	4.79×10^5	1.00×10^6	2.28×10^5	1.09×10^5	$I_{\text{M}} =$	3.26×10^8			
7	313	8.29×10^5	8.74×10^5	1.72×10^6	4.43×10^5	2.24×10^5					

Columns A and B contain the experimental data points for the TL glow curve, while columns C–F contain the calculated data points using equation (2.2) for four values of the energy parameter E ($E = 0.9$, 1.0, 1.1, and 1.2 eV).

The following expression is used to calculate the values of the fitted data in column C, using equation (2.2) for first-order kinetics:

Cell C3 = \$H\$6*EXP(1+\$H\$2/(0.00008617*A3)

*((A3-\$H\$5)/\$H\$5)-((A3*A3)/(\$H\$5*\$H\$5))

*(1-2*0.00008617*\$H\$5/\$H\$2)*EXP(\$H\$2/(0.00008617*A3)

*((A3-\$H\$5)/\$H\$5))-2*0.00008617*\$H\$5/\$H\$2).

This expression refers to cell A3 which contains the absolute temperature T(K). Also, note that cell H2 in the spreadsheet contains the value of the energy parameter $E = 1.0$ eV, cell H5 contains the value of the experimental parameter $T_{\text{M}} = 384$ K, and cell H6 contains the value of the experimental maximum height parameter $I_{\text{M}} = 3.26 \times 10^8$. The above spreadsheet expression refers to the values contained in these cells by using the Excel commands \$H\$2, \$H\$5, \$H\$6, correspondingly.

The user controls the value of the parameter E by changing the value in cell H2, and the whole spreadsheet calculation is automatically updated.

The graphs calculated for $E = 0.9$, 1.0, 1.1, and 1.2 eV are shown in Figure 2.8.

It can be seen that when the value of E is too low (graph corresponding to $E = 0.9$ eV), the calculated TL points lie above the experimental data. This is also evident by inspection of the calculated $I(T)$ values in Table 2.4. On the other hand, when the value of E is too high (graphs corresponding to $E = 1.1$ and 1.2 eV), the calculated TL points lie clearly below the experimental data.

This procedure is a simple example of a glow-curve fitting procedure, in which we find the value of E that yields the best fit to experimental data obeying first-order kinetics.

A more precise numerical method of performing the same fitting procedure is by calculating the FOM for the graphs above. The FOM is defined as [2]

$$\text{FOM} = \frac{\sum_{p} |y_{\text{experimental}} - y_{\text{fit}}|}{\sum_{p} y_{\text{fit}}}, \tag{2.3}$$

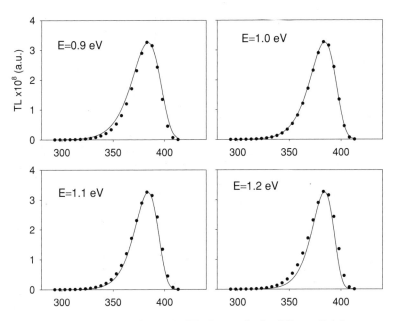

FIGURE 2.8. Calculated first-order TL glow peaks for different E-values.

where $y_{\text{experimental}}$ and y_{fit} represent the experimental TL intensity data and the values of the fitting function, respectively. The summation extends over all the available experimental points.

Table 2.5 shows an example of a FOM calculation as applied to the previous data. Column A contains the experimental data points and columns B and C contain the calculated data points using equation (2.2) for first-order kinetics, for two values of the energy parameter E ($E = 1.0$ and 0.9 eV).

Columns E and F contain the calculation of the quantity $|\text{TL}_{\text{experimental}} - \text{TL}_{\text{calculated}}|$, and cells E29 and F29 contain the calculated values of the FOM. The expressions used in this example are

$$\text{Cell E3} = \text{ABS(A3} - \text{B3)}$$
$$\text{Cell F3} = \text{ABS(A3} - \text{C3)}$$
$$\text{Cell E29} = \text{SUM(E3:E27)/SUM(B3:B27)}$$
$$\text{Cell F29} = \text{SUM(F3:F27)/SUM(C3:C27)}.$$

The FOM for the value of the parameter $E = 1.0$ eV is equal to $0.026 = 2.6\%$, almost four times smaller than the FOM $= 0.094 = 9.4\%$ for the case $E = 0.9$ eV.

The frequency factor s can be calculated by using the value of $E = 1.0$ eV and the temperature of maximum TL intensity $T_{\text{M}} = 384$ K in equation (1.8) for first-order kinetics:

$$s = \frac{\beta E}{kT_{\text{M}}^2} \exp\left(\frac{E}{kT_{\text{M}}}\right) = \frac{(1)1}{(8.617 \times 10^{-5})(384)^2} \exp\left(\frac{1}{(8.617 \times 10^{-5})384}\right)$$
$$= 1.05 \times 10^{12}\, \text{s}^{-1}$$

TABLE 2.5. Example of a FOM calculation for first-order glow curve

	A	B	C	D	E	F
1	TL$_{\text{experimental}}$	$I(T)$ $E = 1$ eV	$I(T)$ $E = 0.9$ eV		$|\text{TL}_{\text{experimental}} - I(t)|$ $E = 1$ eV	$|\text{TL}_{\text{experimental}} - I(t)|$ $E = 0.9$ eV
2						
3	7.56×10^4	6.96×10^4	1.77×10^5		6.01×10^3	1.07×10^5
4	1.28×10^5	1.35×10^5	3.21×10^5		6.83×10^3	1.86×10^5
5	2.44×10^5	2.57×10^5	5.72×10^5		1.30×10^4	3.15×10^5
6	4.54×10^5	4.79×10^5	1.00×10^6		2.42×10^4	5.23×10^5
7	8.29×10^5	8.74×10^5	1.72×10^6		4.41×10^4	8.46×10^5
8	1.49×10^6	1.56×10^6	2.90×10^6		7.88×10^4	1.34×10^6
9	2.61×10^6	2.75×10^6	4.82×10^6		1.38×10^5	2.07×10^6
10	4.51×10^6	4.75×10^6	7.87×10^6		2.38×10^5	3.13×10^6
11	7.65×10^6	8.05×10^6	1.26×10^7		4.02×10^5	4.59×10^6
12	1.27×10^7	1.34×10^7	2.00×10^7		6.64×10^5	6.56×10^6
13	2.09×10^7	2.19×10^7	3.10×10^7		1.07×10^6	9.05×10^6
14	3.35×10^7	3.52×10^7	4.72×10^7		1.68×10^6	1.20×10^7
15	5.27×10^7	5.52×10^7	7.03×10^7		2.55×10^6	1.51×10^7
16	8.07×10^7	8.44×10^7	1.02×10^8		3.67×10^6	1.77×10^7
17	1.20×10^8	1.25×10^8	1.44×10^8		4.93×10^6	1.89×10^7
18	1.71×10^8	1.77×10^8	1.94×10^8		5.92×10^6	1.74×10^7
19	2.31×10^8	2.37×10^8	2.49×10^8		5.84×10^6	1.25×10^7
20	2.90×10^8	2.93×10^8	2.98×10^8		3.62×10^6	5.24×10^6
21	3.26×10^8	3.25×10^8	3.25×10^8		1.16×10^6	1.79×10^6
22	3.15×10^8	3.08×10^8	3.12×10^8		6.59×10^6	3.52×10^6
23	2.43×10^8	2.35×10^8	2.51×10^8		7.99×10^6	1.66×10^7
24	1.34×10^8	1.32×10^8	1.61×10^8		2.81×10^6	2.91×10^7
25	4.49×10^7	4.79×10^7	7.50×10^7		3.03×10^6	2.70×10^7
26	6.75×10^6	9.60×10^6	2.29×10^7		2.85×10^6	1.34×10^7
27	2.57×10^5	8.39×10^5	4.00×10^6		5.83×10^5	3.16×10^6
28						
29				FOM =	0.026	0.094

The resolution of the Kitis et al fitting method can be refined by repeating this process of calculating the FOM for different values of E spaced much closer together (e.g. $E = 1.01$, 1.00, 0.99, etc.) and finding the value of E that minimizes the value of the FOM.

Finally, we summarize in Table 2.6 the results of the various methods for analyzing the given experimental data.

TABLE 2.6. Summary of the results of various analysis methods for first-order glow curve

	E(eV)	s(s^{-1})	Comments below
Initial rise method	0.986 ± 0.003		[1, 5]
Chen's τ-method	0.956		[2, 5]
Chen's δ-method	1.03 ± 0.18		[2, 5]
Chen's ω-method	0.996		[2, 5]
Whole glow-peak method	0.979 ± 0.003	$(4.7 \pm 0.5) \times 10^{10}$	[3]
Fitting method using Kitis et al equation (equation (2.21))	1.1 ± 0.1	1.05×10^{12}	[4, 5]

Comments on the Results of Exercise 2.1

1. The value of E obtained from the IR method is independent of the kinetics of the TL glow peak.

   The presence of thermal quenching affects the value of E obtained in the IR method. A possible correction method for the value of E is given in Chapter 5.

   It is best to use the IR method with samples irradiated at low doses, i.e. samples away from saturation conditions [3].

   The test dose used to obtain the TL glow curve is our "probe" of the material, and must always be as small as possible, so that on one hand it does not disturb the system and on the other it can give us a statistically satisfactory signal. Typical values of test doses may be in the mGy or µGy range.

2. The value of E obtained with peak shape methods can be influenced by the presence of smaller satellite peaks.

3. The whole glow-curve method yields information on both E and the pre-exponential factor s. By using the values of E, s, and n_0 (obtained from the area under the glow curve), it is possible to compare directly the experimental data with the TL intensity obtained using equation (1.5) (see also Exercises 2.4–2.6 in this chapter for second- and general-order kinetics).

4. The Kitis et al method is based on two experimentally measured parameters, T_M and I_M. The activation energy E is treated as a fitting parameter. The method can be easily adopted on a computer to yield high accuracy for E.

5. The pre-exponential factor s can be calculated from the value of T_M, E, and β by using equation (1.8). The estimated uncertainties $\Delta s/s$ from equation (1.8) can be very large (50–100%), even when the uncertainty $\Delta E/E$ is very small.

(e) Can it be concluded for this material that this TL peak will always follow first-order kinetics?

In general, one cannot assume that the studied TL glow curve of this material will always follow first-order kinetics. The analysis should be carried out for glow peaks measured under different heating rates, various irradiation doses, powdered and bulk samples, etc.

Some of the criteria for first-order kinetics are:

I. *Peak shape*: First-order peaks have $\mu = 0.42$.

II. *Peak shift*: In first-order TL glow peaks, the location of maximum TL intensity does not shift in temperature for different irradiation doses.

III. *Isothermal decay results*: These can provide valuable independent information about the kinetics of the TL process involved at different temperatures. First-order kinetics corresponds to exponential isothermal decay curves.

Exercise 2.2: Heating Rate Method for First-Order Kinetics

You are given the data in Figure 2.9 for four experimental TL glow curves measured with different heating rates (TL versus Temperature T, and known heating rates $\beta_1 = 0.5$, $\beta_2 = 1$, and $\beta_3 = 2$, and $\beta_3 = 3\,\mathrm{K\,s^{-1}}$). It is known that this TL peak follows first-order kinetics.

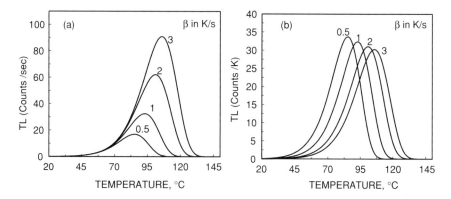

FIGURE 2.9. The experimental TL glow curves for different heating rates. (a) The y-axis is represented as counts/s and (b) the y-axis is represented as counts/K.

(a) Discuss the observed changes in the peak position and peak shape for different heating rates.
(b) Apply the two-heating rate equation for E (equation (1.26)), to obtain a quick estimate of the activation energy E.
(c) Apply also the $I_M - T_M$ variation method (equation (1.31)), to obtain a quick estimate of the activation energy E.
(d) By applying the heating rate method of analysis find the kinetic parameters E and s, and their uncertainties ΔE and Δs.

Solution

(a) The data of Figure 2.9 show that as the heating rate increases, the glow peaks shift to higher temperatures, and the height of the TL peak changes. Because in a typical TL experiment, one collects the TL signal as a function of time, the y-axis in Figure 2.9(a) is represented in counts/s. These units of counts/s are not suitable for graphing the actual TL glow curve which is a function of temperature, so it is necessary to convert into temperature units. This is done by dividing each of the graphs in Figure 2.9(a) by the corresponding heating rate β, and one obtains the y-axis in counts/K as shown in Figure 2.9(b).

The area under the peaks in Figure 2.9(a) is proportional to the heating rate β, whereas the area under the glow curves in Figure 2.9(b) is constant.

The temperatures T_M for the maximum TL intensity and the corresponding intensities I_M are found from the curves of Figure 2.9(a) and (b), and are listed in Table 2.7.

(b) We can calculate the energy E from the two-heating rate equation (equation (1.26))

$$E = k \frac{T_{M1} T_{M2}}{T_{M1} - T_{M2}} \ln \left[\frac{\beta_1}{\beta_2} \left(\frac{T_{M2}}{T_{M1}} \right)^2 \right]. \qquad (1.26)$$

TABLE 2.7. Calculation of $\ln(T_M^2/\beta)$ and $1/kT_M$ for first-order glow curve

β(K s^{-1})	T_M(°C)	T_M(K)	$\ln(T_M^2/\beta)$	$1/kT_M$(eV)$^{-1}$	I_M(counts/s)	I_M(counts/K)
0.5	84	357	12.449	32.507	17	34
1	92	365	11.800	31.794	32	32
2	100	373	11.150	31.113	62	31
3	104	377	10.766	30.782	90	30

Inserting $T_{M1} = 357$ K, $T_{M2} = 365$ K, $\beta_1 = 0.5$ K s^{-1}, $\beta_2 = 1$ K s^{-1}, we obtain

$E = 8.617 \times 10^{-5}$ (357)(365)/(357 − 365) ln[0.5(365)^2/1(357)^2] = 0.911 eV.

(c) We can also estimate the energy E from the two-intensities equation (equation (1.31))

$$E = \frac{kT_{m1}T_{m2}}{T_{m1} - T_{m2}} \ln \frac{I_{m1}}{I_{m2}}. \qquad (1.31)$$

Inserting $T_{M1} = 357$ K, $T_{M2} = 365$ K, $I_{M1} = 34$ (counts/K), $I_{M2} = 32$ (counts/K), we obtain

$E = 8.617 \times 10^{-5}$ (357)(365)/(357 − 365) ln[(34/32)0.5] = 0.89 eV.

(d) We calculate the quantities $1/kT_M$ (k = Boltzmann constant) and $\ln(T_M^2/\beta)$ shown in Table 2.7 with β = given heating rates. As discussed in Chapter 1, equation (1.27) shows that the slope of the graph $\ln(T_M^2/\beta)$ versus $1/kT_M$ will be equal to the activation energy E, and that the y-intercept will be equal to $\ln(E/sk)$.

From the slope and intercept of the graph $\ln(T_M^2/\beta)$ versus $1/kT_M$, in Figure 2.10, we can calculate the kinetic parameters E and s as follows:

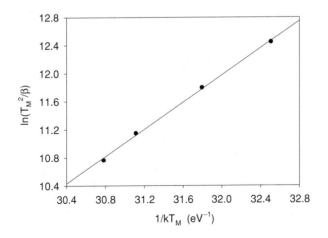

FIGURE 2.10. Graph of $\ln(T_M^2/\beta)$ versus $1/kT_M$ to determine E and s.

From the slope of the graph, $E = 0.9668$ eV.
From the value of the y-intercept $= \ln(E/sk)$, we obtain

$$s = Ee^{\text{intercept}/k} = 0.9668e^{(18.962)/(8.617 \times 10^{-5})} = 1.9 \times 10^{12} \text{ s}^{-1}.$$

The values of E and s obtained above can be checked for self-consistency as follows. Because the temperature T_M of the maximum TL intensity is known from the experimental data, the value of s can be calculated in an independent manner by rearranging equation (1.8) to obtain

$$s = \frac{\beta E}{kT_M^2} \exp\left(\frac{E}{kT_M}\right). \tag{2.4}$$

By using the values of $E = 0.9668$ eV, $T_M = 84°C = 357$ K, $\beta = 0.5$ K s^{-1} in equation (2.4):

$$s = \frac{(0.5)(0.9668)}{(8.617 \times 10^{-5})(357)^2} \exp\left(\frac{0.9668}{(8.617 \times 10^{-5})357}\right) = 2.1 \times 10^{12} \text{ s}^{-1}.$$

This value of s is very close to the value of $s = 1.9 \times 10^{12}$ s^{-1} obtained above using the y-intercept of the graph, indicating that the results of the heating rate method are self-consistent with the assumption of first-order kinetics.

(d) The errors ΔE and Δs can be calculated from the uncertainties in the slope and y-intercept of the best-fitting regression lines.

From the slope of the regression line, $E = 0.967 \pm 0.029$ eV.

This corresponds to a percent error in E of $100(\Delta E/E) = 100(0.029/0.967) = 3\%$.

By taking the logarithmic derivative of the equation $s = E \exp$ (intercept)/k, we obtain

$$\ln s = \ln E + \text{intercept} - \ln k,$$

$$\left|\frac{\Delta s}{s}\right| = \left|\frac{\Delta E}{E}\right| + |\Delta(\text{intercept})| = \left|\frac{0.0288}{0.9668}\right| + |0.91| = 0.94. \tag{2.5}$$

This leads to a very large (but nevertheless typical) uncertainty in s, of the order of 94%.

As a general comment on applying the heating rate methods of analysis, we wish to point out that the methods based on the variation of I_M with the heating rate β are easy to use. These methods are perhaps also more reliable than the methods based on the changes of T_M with the heating rate, because they would be less affected by the presence of nearby overlapping TL peaks. It is rather strange that these I_M-based methods have not been very popular in the TL literature. In our opinion, this is due most probably to confusion between the theoretical heights I_M and the corresponding heights measured in an experiment. These latter heights must be divided by the heating rate β as was shown in this exercise, in order to correct the units and to normalize the areas under the TL glow peaks.

When using theoretical methods involving I_M, the units of the height I_M are in counts/s. However, in experimental data, one measures I_M in units of counts/K.

If one tries to apply the I_M methods of analysis using the experimental data in counts/K, these methods fail dramatically to yield the correct values of E. For correct application of the method, the experimental values of counts/K must be changed into counts/s by multiplying with the heating rate β.

This important point has not been emphasized or clarified enough in the TL literature.

Exercise 2.3: Isothermal Method for First-Order Kinetics

You are given the experimental data in Table 2.8 for the isothermal decay curves of a TL peak, which were measured for four different temperatures of $T = 50°C, 60°C, 70°C,$ and $80°C$.

(a) Show that these data are consistent with the assumption that the TL glow peak follows first-order kinetics.
(b) Find the kinetic parameters E and s.

Solution

(a) Figure 2.11 shows the given data for the four different temperatures $T = 50°C, 60°C, 70°C,$ and $80°C$.

As discussed in Chapter 1, the isothermal decay curves for TL peaks following first-order kinetics are exponential functions of time, given by

$$I = I_0 \exp(-s \cdot \exp(-E/kT) \cdot t). \qquad (2.6)$$

This equation tells us that a graph of $\ln(I)$ versus time t will be linear for first-order kinetics peaks, and that the slope of the line will be

$$|\text{slope}| = s \cdot \exp(-E/kT). \qquad (2.7)$$

By taking the natural logarithm of this equation, we obtain

$$\ln(|\text{slope}|) = \ln s - E/kT. \qquad (2.8)$$

TABLE 2.8. Data for the isothermal decay curves of a first-order TL peak

$t(s)$	TL, $T = 50°C$	TL, $T = 60°C$	TL, $T = 70°C$	TL, $T = 80°C$
0	2.48×10^6	7.22×10^6	1.94×10^7	4.74×10^7
20	2.47×10^6	7.11×10^6	1.87×10^7	4.26×10^7
40	2.45×10^6	7.01×10^6	1.79×10^7	3.83×10^7
60	2.44×10^6	6.91×10^6	1.72×10^7	3.45×10^7
80	2.43×10^6	6.80×10^6	1.65×10^7	3.10×10^7
100	2.42×10^6	6.70×10^6	1.58×10^7	2.78×10^7
120	2.40×10^6	6.61×10^6	1.52×10^7	2.50×10^7
140	2.39×10^6	6.51×10^6	1.46×10^7	2.25×10^7
160	2.38×10^6	6.41×10^6	1.40×10^7	2.02×10^7
180	2.37×10^6	6.32×10^6	1.34×10^7	1.82×10^7
200	2.36×10^6	6.23×10^6	1.29×10^7	1.63×10^7
220	2.35×10^6	6.13×10^6	1.24×10^7	1.47×10^7

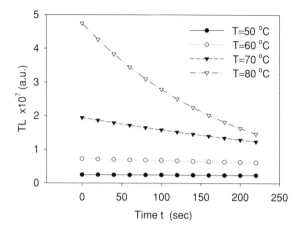

FIGURE 2.11. The isothermal decay curves for first-order TL data.

The graph of the $\ln(|\text{slope}|)$ versus $1/kT$ should be a straight line with slope $= -E$ and y-intercept $= \ln s$.

We first find $\ln(\text{TL})$ for each isothermal curve as shown in Table 2.9.

By graphing the $\ln(\text{TL})$ versus time, we obtain straight lines as shown in Figure 2.12, indicating that the given isothermal TL data obey first-order kinetics. We next find the regression lines through each of the graphs shown in Figure 2.12.

(b) Next, we tabulate in Table 2.10 the slopes of these linear graphs and calculate the natural logarithm of the slopes, $\ln(|\text{slope}|)$. Finally, we graph in Figure 2.13 the $\ln(|\text{slope}|)$ versus $1/kT$, where $T =$ temperature (in K) at which the isothermal decay curves were measured.

The slope of the regression line gives the activation energy E:

$$E = 1.007 \pm 0.002 \text{ eV}.$$

TABLE 2.9. The $\ln(\text{TL})$ data for each isothermal curve

$t(s)$	TL, $T = 50°C$	$\ln(\text{TL})$, $T = 50°C$	TL, $T = 60°C$	$\ln(\text{TL})$, $T = 60°C$	TL, $T = 70°C$	$\ln(\text{TL})$, $T = 70°C$	TL, $T = 80°C$	$\ln(\text{TL})$, $T = 8$
0	2.48×10^6	14.726	7.22×10^6	15.796	1.94×10^7	16.783	4.74×10^7	17.68.
20	2.47×10^6	14.722	7.11×10^6	15.782	1.87×10^7	16.750	4.26×10^7	17.594
40	2.45×10^6	14.718	7.01×10^6	15.771	1.79×10^7	16.718	3.83×10^7	17.498
60	2.44×10^6	14.712	6.91×10^6	15.749	1.72×10^7	16.674	3.45×10^7	17.386
80	2.43×10^6	14.705	6.80×10^6	15.735	1.65×10^7	16.629	3.10×10^7	17.264
100	2.42×10^6	14.701	6.70×10^6	15.720	1.58×10^7	16.584	2.78×10^7	17.15(
120	2.40×10^6	14.696	6.61×10^6	15.707	1.52×10^7	16.548	2.50×10^7	17.05'
140	2.39×10^6	14.690	6.51×10^6	15.699	1.46×10^7	16.511	2.25×10^7	16.96.
160	2.38×10^6	14.687	6.41×10^6	15.683	1.40×10^7	16.462	2.02×10^7	16.86.
180	2.37×10^6	14.681	6.32×10^6	15.660	1.34×10^7	16.420	1.82×10^7	16.754
200	2.36×10^6	14.675	6.23×10^6	15.645	1.29×10^7	16.385	1.63×10^7	16.634
220	2.35×10^6	14.673	6.13×10^6	15.632	1.24×10^7	16.346	1.47×10^7	16.52.

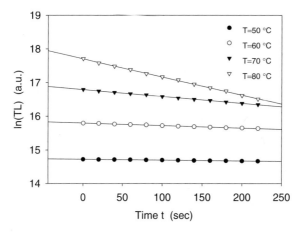

FIGURE 2.12. The isothermal decay curves on semilog scale for first-order TL data.

And the frequency factor s can be found from the intercept of the regression line:

$$\text{Intercept} = \ln(s) = 27.87 \pm 0.07.$$

Therefore,

$$s = \exp(27.87) = 1.3 \times 10^{12} \text{ s}^{-1}.$$

The error in the frequency factor Δs can be calculated from the uncertainties in the intercept of the regression line as follows:

$$\Delta(\text{intercept}) = \Delta s / s = 0.07.$$

This gives a rather unusually small error of 7% for the value of the frequency factor s.

Self-Consistency Check of E and s Values

The values of s and E can be checked for self-consistency as follows: We can calculate theoretical slopes of the isothermal decay curves using the E and s values and compare them with the experimental slopes obtained from the graphs. Theoretically, the slopes of the graphs $\ln(\text{TL})$ versus time t should be given by equation (2.7)

$$\text{slope} = s \exp\left(\frac{-E}{kT}\right).$$

TABLE 2.10. The slopes of linear isothermal graphs

$T(°C)$	$|\text{slope}|(\text{s}^{-1})$	$1/kT$ (eV^{-1})	$\ln(|\text{Slope}|)$	Calculated $|\text{slope}|(\text{s}^{-1})$	%Difference in slopes
50	2.46×10^{-4}	35.93	-8.31	2.49×10^{-4}	1.2
60	7.35×10^{-4}	34.85	-7.22	7.33×10^{-4}	-0.3
70	2.04×10^{-3}	33.83	-6.19	2.02×10^{-3}	-0.8
80	5.33×10^{-3}	32.88	-5.23	5.28×10^{-3}	-1.0

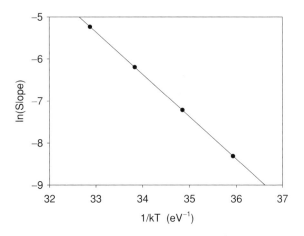

FIGURE 2.13. The ln(|slope|) versus $1/kT$ graph for first-order TL data.

By using the values of $E = 1.007$ eV and $s = 1.3 \times 10^{12}$ s^{-1} in equation (2.7), we obtain the theoretical values of the slopes shown on the fifth column of Table 2.10. The last column in Table 2.10 shows that the percent difference between the theoretical and experimental slopes in columns 2 and 5 is very small, of the order of 1%, indicating that the isothermal decay data are consistent with first-order kinetics.

Exercise 2.4: Analysis of a Second-Order TL Peak

You are given in Table 2.11 and Figure 2.14 the experimental data for a TL glow curve (TL versus Temperature T), which was measured with a heating rate $\beta = 1$ K s^{-1}.

(a) Apply the IR method to find the activation energy E. The value for E obtained using the IR method is assumed to be independent of the order of kinetics.

TABLE 2.11. The experimental data for a second-order TL glow curve

$T(^{\circ}C)$	TL (a.u.)	$T(^{\circ}C)$	TL (a.u.)
46	1.58×10^6	124	1.58×10^8
52	3.09×10^6	130	1.20×10^8
58	5.87×10^6	136	8.64×10^7
64	1.09×10^7	142	5.98×10^7
70	1.95×10^7	148	4.05×10^7
76	3.37×10^7	154	2.71×10^7
82	5.60×10^7	160	1.81×10^7
88	8.78×10^7	166	1.21×10^7
94	1.28×10^8	172	8.17×10^6
100	1.70×10^8	178	5.53×10^6
106	2.00×10^8	184	3.77×10^6
112	2.08×10^8	190	2.59×10^6
118	1.92×10^8	196	1.79×10^6

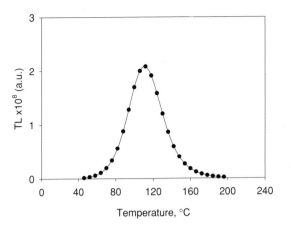

FIGURE 2.14. The second-order TL glow curve.

(b) Apply Chen's peak shape equations for second-order kinetics to find the activation energy E, using the shape parameters τ, δ, and ω. By assuming that the experimental error in the quantities τ, δ, and ω is $\Delta T = 2$ K, estimate the error $\Delta\mu$ in the value of the geometrical shape factor μ.

Show that the values of μ and $\Delta\mu$ are consistent with the assumption that the TL glow curve obeys second-order kinetics.

(c) Apply the whole glow-peak method to the data given and find E, s, and the order of kinetics b. Verify that the given TL glow curve corresponds to second-order kinetics.

(d) Using the values of I_M (maximum TL intensity) and T_M (temperature of maximum intensity) from the data table, do a curve fitting to the given numerical data. Use the following analytical equation developed by Kitis et al [2] for second-order kinetics:

$$I(T) = 4I_M \exp\left(\frac{E}{kT} \cdot \frac{T - T_M}{T_M}\right)$$

$$\times \left[\frac{T^2}{T_M^2} \cdot \left(1 - \frac{2kT}{E}\right) \exp\left(\frac{E}{kT} \cdot \frac{T - T_M}{T_M}\right) + 1 + \frac{2kT_M}{E}\right]^{-2}. \quad (2.9)$$

The activation parameter E in this expression can be treated as an adjustable parameter.

Graph both the experimental data and the calculated second-order TL glow curve on the same graph and compare them.

Calculate the FOM for the TL glow curve.

(e) Can it be concluded from the above analysis that this material will always follow second-order kinetics?

Solution

(a) *The IR method.* We calculate in Table 2.12 the values of $1/kT$ (T = temperature in K) and the values of ln(TL) in a spreadsheet.

TABLE 2.12. Calculated values of $1/kT$ and the values of ln(TL)

$T(°C)$	$TL_{experimental}$	$1/kT$ (eV^{-1})	ln(TL)	$T(°C)$	$TL_{experimental}$	$1/kT$ (eV^{-1})	ln(TL)
46	1.58×10^6	36.38	14.27	124	1.58×10^8	29.23	18.88
52	3.09×10^6	35.71	14.94	130	1.20×10^8	28.80	18.61
58	5.87×10^6	35.06	15.59	136	8.64×10^7	28.37	18.27
64	1.09×10^7	34.44	16.20	142	5.98×10^7	27.96	17.91
70	1.95×10^7	33.83	16.78	148	4.05×10^7	27.57	17.52
76	3.37×10^7	33.25	17.33	154	2.71×10^7	27.18	17.12
82	5.60×10^7	32.69	17.84	160	1.81×10^7	26.80	16.71
88	8.78×10^7	32.15	18.29	166	1.21×10^7	26.44	16.31
94	1.28×10^8	31.62	18.67	172	8.17×10^6	26.08	15.92
100	1.70×10^8	31.11	18.95	178	5.53×10^6	25.73	15.53
106	2.00×10^8	30.62	19.11	184	3.77×10^6	25.39	15.14
112	2.08×10^8	30.14	19.15	190	2.59×10^6	25.06	14.77
118	1.92×10^8	29.68	19.07	196	1.79×10^6	24.74	14.40

We next graph in Figure 2.15(a) the ln(TL) versus $1/kT$ data and find a regression line through the first 7 data points, as shown in Figure 2.15(b).

The slope of the regression line gives the activation energy E as

$$E = 0.969 \pm 0.006 \text{ eV}, \qquad \text{with } R^2 = 0.9997.$$

(b) *Chen's peak shape equations.* From the given experimental data, we can estimate the temperatures

$$T_1 = 91°C = 364 \text{ K}, \quad T_2 = 133°C = 406 \text{ K}, \quad T_M = 112°C = 385 \text{ K},$$

where

T_M = peaktemperature at the maximum TL intensity,

T_1, T_2 = temperatures on either side of T_M, corresponding to the half-maximum intensity.

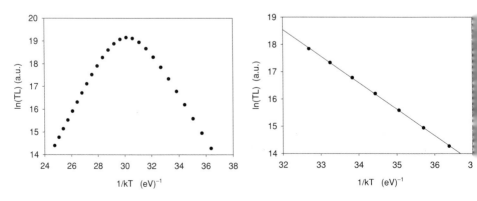

FIGURE 2.15. IR method analysis.

We next calculate the quantities μ, τ, δ, and ω:

$$\tau = T_M - T_1 = 21 \text{ K}, \quad \delta = T_2 - T_M = 21 \text{ K}, \quad \omega = T_2 - T_1 = 42 \text{ K},$$
$$\mu = \delta/\omega = 21/42 = 0.50.$$

The calculated value of the geometrical shape factor $\mu = \delta/\omega = 0.50$ is very close to the theoretical value for a second-order TL peak $\mu = \delta/\omega = 0.52$.

Using the known experimental error $\Delta T = 2 \text{ K}$ for the quantities τ, δ, and ω, we can do an error analysis of the values of μ. As in the case of first-order kinetics,

$$\left| \frac{\Delta \mu}{\mu} \right| = \left| \frac{\Delta \delta}{\delta} \right| + \left| \frac{\Delta \omega}{\omega} \right| = \left| \frac{2}{21} \right| + \left| \frac{2}{42} \right| = 0.095 + 0.048 = 0.143.$$

This leads to a value of $\mu \pm \Delta \mu = 0.50 \pm 0.07$, which is consistent with second-order kinetics within the accuracy of the given TL data.

We apply Chen's equation for second-order kinetics.

Using the value of τ:

$$E = \frac{1.81 k T_M^2}{\tau} - 2(2kT_M) = 1.101 - 0.133 = 0.968 \text{ eV}.$$

Using the value of δ:

$$E = \frac{1.71 k T_M^2}{\delta} = 1.040 \text{ eV}.$$

Using the value of ω:

$$E = \frac{3.54 k T_M^2}{\omega} - 2kT_M = 1.077 - 0.066 = 1.011 \text{ eV}.$$

In order to find the error ΔE in the activation energy E, we take the logarithmic derivative of the equation $E = 1.71 k T_M^2/\delta$:

$$\left| \frac{\Delta E}{E} \right| = 2 \left| \frac{\Delta T_M}{T_M} \right| + \left| \frac{\Delta \delta}{\delta} \right| = 2 \left| \frac{2}{385} \right| + \left| \frac{2}{21} \right| = 0.010 + 0.095 = 0.105.$$

This gives a rather large 10.5% error of $\Delta E = 0.105E = 0.105(1.040) = 0.11 \text{ eV}$.

(c) *The whole glow-peak method.* We graph $\ln(I/n^b)$ versus $1/T$ for various values of b between 1.8 and 2.1, and find the value of b that gives a linear graph. As in the case of first-order kinetics, $n(T)$ is the area under the glow peak and it is calculated starting at a temperature T, up to the maximum temperature available in the experimental data. In the data shown in Table 2.13, the maximum available temperature is $196°C$.

By following the same procedure as in the case of first-order kinetics, we set up a spreadsheet to calculate the quantities $\ln(I/n^b)$ and $1/kT$ as shown in Table 2.13.

Additional columns are created in the spreadsheet for the quantities of $\ln(TL/n^b)$ for several values of the kinetic order $b = 2.0$, 2.1, 1.9, and 1.8.

Finally, several graphs are drawn of $\ln(TL/Area^b)$ versus $1/kT$ as shown in Figure 2.16.

It is clear that all four graphs in Figure 2.16 deviate from straight lines, especially at low values of $1/kT$ (which correspond to large temperatures located on the high-temperature end of the TL glow peak). These deviations are due to experimental

TABLE 2.13. The quantities $\ln(I/n^b)$ and $1/kT$

		A	B	C	D	E	F	G
l	$T(°C)$	$1/kT$ $(\text{eV})^{-1}$	Area	$\ln(\text{TL/Area})$	$\ln(\text{TL}/n^2)$	$\ln(\text{TL}/n^{2.1})$	$\ln(\text{TL}/n^{1.9})$	$\ln(\text{TL}/n^{1.8})$
2	46	36.38	9.96×10^9	-8.75	-31.77	-34.07	-29.47	-27.17
3	52	35.71	9.95×10^9	-8.08	-31.10	-33.40	-28.80	-26.49
4	58	35.06	9.94×10^9	-7.43	-30.45	-32.75	-28.15	-25.85
5	64	34.44	9.90×10^9	-6.82	-29.83	-32.13	-27.53	-25.23
6	70	33.83	9.84×10^9	-6.22	-29.23	-31.53	-26.93	-24.63
7	76	33.25	9.72×10^9	-5.66	-28.66	-30.96	-26.36	-24.06
8	82	32.69	9.52×10^9	-5.14	-28.11	-30.41	-25.81	-23.52
9	88	32.15	9.18×10^9	-4.65	-27.59	-29.88	-25.30	-23.00
10	94	31.62	8.65×10^9	-4.21	-27.10	-29.38	-24.81	-22.52
11	100	31.11	7.89×10^9	-3.84	-26.63	-28.91	-24.35	-22.07
12	106	30.62	6.87×10^9	-3.54	-26.19	-28.45	-23.92	-21.66
13	112	30.14	5.67×10^9	-3.30	-25.76	-28.01	-23.52	-21.21
14	118	29.68	4.42×10^9	-3.14	-25.35	-27.57	-23.13	-20.91
15	124	29.23	3.27×10^9	-3.03	-24.94	-27.13	-22.74	-20.55
16	130	28.80	2.32×10^9	-2.96	-24.52	-26.68	-22.37	-20.21
17	136	28.37	1.60×10^9	-2.92	-24.11	-26.23	-21.99	-19.87
18	142	27.96	1.08×10^9	-2.89	-23.69	-25.77	-21.61	-19.53
19	148	27.57	7.19×10^8	-2.88	-23.27	-25.31	-21.23	-19.19
20	154	27.18	4.76×10^8	-2.86	-22.84	-24.84	-20.85	-18.85
21	160	26.80	3.13×10^8	-2.85	-22.41	-24.36	-20.45	-18.50
22	166	26.44	2.04×10^8	-2.82	-21.95	-23.87	-20.04	-18.13
23	172	26.08	1.31×10^8	-2.78	-21.47	-23.34	-19.60	-17.73
24	178	25.73	8.21×10^7	-2.70	-20.92	-22.74	-19.10	-17.28
25	184	25.39	4.89×10^7	-2.56	-20.27	-22.04	-18.50	-16.73
26	190	25.06	2.63×10^7	-2.32	-19.40	-21.11	-17.69	-15.99
27	196							

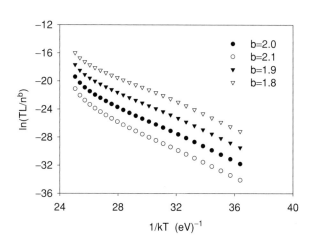

FIGURE 2.16. Graphs of $\ln(\text{TL/Area}^b)$ versus $1/kT$ for several values of kinetic order b.

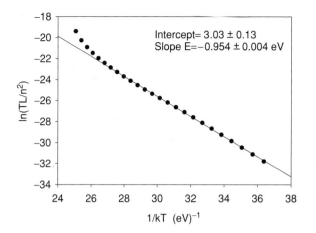

FIGURE 2.17. The parameters E and s' can be calculated from the whole glow-peak method.

uncertainties in the data, and also to the fact that only 26 data points are available on the TL glow curve.

A regression line is fitted to the four graphs above:

$$b = 1.8: \quad R^2 = 0.9983$$
$$b = 1.9: \quad R^2 = 0.9992$$
$$b = 2.0: \quad R^2 = 0.9990$$
$$b = 2.1: \quad R^2 = 0.9979.$$

The regression line for $b = 1.9$ gives the largest regression coefficient R^2, with a value that is very close to the regression coefficient for the case $b = 2.0$. Within the accuracy of the given experimental data and within the framework of the whole glow-peak method of analysis, we can conclude that the given TL glow peak data follow second-order kinetics.

The values of E and s' can be calculated from the best-fitting regression line shown in Figure 2.17:

$$\text{Best intercept} = 3.03 \pm 0.13,$$
$$\text{Best slope } E = -0.954 \pm 0.004 \text{ eV}.$$

According to equation (1.23), the value of s' can be calculated from the intercept of the regression line:

$$s' = \beta e^{(\text{intercept})} = 1 e^{(3.03)} = 20.697.$$

The whole glow-peak method yields information about both the activation energy E and the effective frequency factor $s' = s/N$ appearing in equation (1.6):

$$I(T) = n_0^2 \frac{s}{N} \exp\left(-\frac{E}{kT}\right) \left[1 + \frac{n_0 s}{\beta N} \int_{T_0}^{T} \exp\left(-\frac{E}{kT'}\right) dT'\right]^{-2}. \quad (1.6)$$

Since TL data analysis of a glow curve cannot yield a value for the *absolute* concentration n_0 of traps in the material, the factors s/N and n_0 appearing in equation (1.6) represent *two empirical fitting parameters* for second-order glow peaks. The value of n_0 can be obtained from the area under the glow curve as follows:

$$\frac{\text{Area}}{\beta} = \frac{1}{\beta} \int_{T_0}^{T_f} I \, dT = \int_{t_0}^{t_f} I \, dt = \int_{t_0}^{t_f} -\left(\frac{dn}{dt}\right) dt$$

$$= n(t_0) - n(t_f) = n_0 - 0 = n_0. \tag{2.10}$$

In our example, the area can be estimated by summing the TL intensities multiplied by the temperature interval $\Delta T = 6$ K between TL measurements, and dividing by the heating rate $\beta = 1$ K s^{-1}:

$$n_0 \approx \frac{1}{\beta} \int_{T_0}^{T_f} I \, dT = \frac{1}{\beta} \sum I(T) \Delta T = \frac{1}{1} \sum I(T)(6K) = 9.96 \times 10^9. \tag{2.11}$$

By using the values of $E = 0.954$ eV, $s' = 20.697$, $n_0 = \text{Area} = 9.96 \times 10^9$, and $\beta = 1$ K s^{-1}, it is possible to calculate the TL intensity using equation (1.6), and to compare this result directly with the given experimental data. The integral in equation (1.6) can be calculated using numerical integration methods, as shown for example in Chapter 3. As an alternative method, the series approximation given in equation (1.52) can be used to evaluate the integral.

The result of the comparison is shown in Figure 2.18, where the calculated $I(T)$ from equation (1.6) is compared with the original experimental data. Figure 2.18 shows that the calculated parameters E, s', and n_0 from the whole glow-peak method, as well as the second-order equation (equation (1.6)), describe the given experimental data in a satisfactory manner.

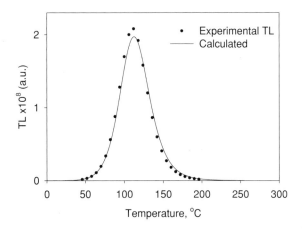

FIGURE 2.18. Comparison of calculated TL intensity using equation (1.6) (solid line), and original experimental data (individual data points). The parameters used in equation (1.6) were calculated using the whole glow-peak method.

TABLE 2.14. Calculations for glow-curve fitting for several E values

A	B	C	D	E	F	G	H	I	J	K	L
		$I(T)$	$I(T)$	$I(T)$	$I(T)$						
T(K)	TL$_{\text{experimental}}$	$E = 1$ eV	$E = 0.9$ eV	$E = 1.1$ eV	$E = 1.2$ eV						
						$E =$	1	eV	0.9	1.1	1.2
319	1.58×10^6	1.49×10^6	2.63×10^6	8.45×10^5	4.76×10^5						
325	3.09×10^6	3.02×10^6	4.97×10^6	1.83×10^6	1.10×10^6						
331	5.87×10^6	5.94×10^6	9.11×10^6	3.85×10^6	2.48×10^6	$T_M =$	380	K			
337	1.09×10^7	1.13×10^7	1.62×10^7	7.83×10^6	5.39×10^6	$I_M =$	2.08×10^8				
343	1.95×10^7	2.08×10^7	2.78×10^7	1.54×10^7	1.13×10^7						

The observed discrepancies between experiment and calculation in Figure 2.18 are due to the several approximations involved in applying the whole glow-peak method, and to the approximation of the area using equation (2.11).

(d) *Glow-curve fitting using the Kitis et al equation.* We use the analytical equation developed by Kitis et al [2], which relies on two experimentally measured quantities, $I_M = 2.08 \times 10^8$ and $T_M = 380$ K:

$$I(T) = 4I_M \exp\left(\frac{E}{kT} \cdot \frac{T - T_M}{T_M}\right)$$

$$\times \left[\frac{T^2}{T_M^2} \cdot \left(1 - \frac{2kT}{E}\right) \exp\left(\frac{E}{kT} \cdot \frac{T - T_M}{T_M}\right) + 1 + \frac{2kT_M}{E}\right]^{-2}. \quad (2.12)$$

The activation parameter E is treated in this equation as an adjustable parameter. We calculate several graphs with values of $E = 0.9$, 1.0, 1.1, and 1.2 eV. The calculations can be easily set up in a spreadsheet as shown in Table 2.14. Only the first 5 rows are shown for the sake of brevity.

Columns A and B contain the experimental data points for the TL glow curve, whereas columns C–F contain the calculated data points using equation (2.12) for second-order kinetics and for four values of the energy parameter E ($E = 0.9$, 1.0, 1.1, and 1.2 eV).

The following equation is used to calculate the values of the fitted data in column C, using equation (2.12) for second-order kinetics:

Cell C3 = 4*$H$6*EXP($H$2/(0.00008617*B3)*((B3-$H$5)/

$H$5))*((B3*B3)/($H$5*$H$5))*((1-2*0.00008617*

$H$5/$H$2)*EXP($H$2/(0.00008617*B3)*((B3-$H$5)/

$H$5))+1+2*0.00008617*$H$5/$H$2)^-2.

Note that cell H2 in the spreadsheet contains the value of the energy parameter $E = 1.0$ eV, cell H5 contains the value of the experimental parameter $T_M = 380$ K, and cell H6 contains the value of the experimental maximum height parameter $I_M = 2.08 \times 10^8$. The above spreadsheet expression refers to the values contained in these cells by using the Excel expressions $H$2, $H$5, $H$6, correspondingly.

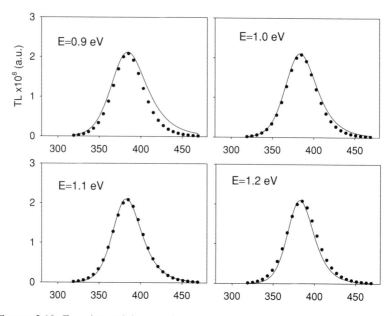

FIGURE 2.19. Experimental data and fitted graphs calculated for several values of E.

Similar expressions are entered in the columns D, E, and F, and the Excel command **Fill Down** is used to fill the rest of these columns.

The graphs calculated for $E = 0.9$, 1.0, 1.1, and 1.2 eV are shown in Figure 2.19, together with the given experimental data.

It can be seen in Figure 2.19 that when the value of E is too low (graph corresponding to $E = 0.9$ eV), the calculated TL points lie well above the experimental data. This is also evident by inspection of the $I(T)$ data in Table 2.14. On the other hand, when the value of E is too high (graph corresponding to $E = 1.2$ eV), the calculated TL points lie below the experimental data.

This procedure is a simple example of a single glow-curve fitting procedure, in which we find the value of E that yields the best fit to experimental data obeying second-order kinetics.

A more precise numerical method of performing a fitting procedure is by calculating the FOM, using a similar calculation to the one employed for first-order kinetics (Exercise 2.1).

Table 2.15 shows an example of a FOM calculation as applied to the previous data. Column A contains the experimental data points and columns B and C contain the calculated data points using equation (2.12) for second-order kinetics and for two values of the energy parameter E ($E = 1.1$ and 0.9 eV).

Columns E and F contain the calculation of the quantity |TL$_{\text{experimental}}$ − TL$_{\text{calculated}}$|, and cells E29 and F29 contain the calculated values of the FOM.

The expressions used in this example are

$$\text{Cell E3} = \text{ABS(A3-B3)}$$
$$\text{Cell F3} = \text{ABS(A3-C3)}$$

TABLE 2.15. Example of a FOM calculation

	A	B	C	D	E	F
		$I(T)$	$I(T)$		$|\text{TL}_{\text{experimental}} - I(t)|$	$|\text{TL}_{\text{experimental}} - I(t)|$
1	$\text{TL}_{\text{experimental}}$	$E = 0.9\,\text{eV}$	$E = 1.1\,\text{eV}$		$E = 0.9\,\text{eV}$	$E = 1.1\,\text{eV}$
2						
3	1.58×10^6	2.63×10^6	8.45×10^5		1.05×10^6	7.38×10^5
4	3.09×10^6	4.97×10^6	1.83×10^6		1.88×10^6	1.26×10^6
5	5.87×10^6	9.11×10^6	3.85×10^6		3.23×10^6	2.03×10^6
6	1.09×10^7	1.62×10^7	7.83×10^6		5.34×10^6	3.03×10^6
7	1.95×10^7	2.78×10^7	1.54×10^7		8.36×10^6	4.10×10^6
8	3.37×10^7	4.59×10^7	2.90×10^7		1.22×10^7	4.77×10^6
9	5.60×10^7	7.20×10^7	5.17×10^7		1.60×10^7	4.24×10^6
10	8.78×10^7	1.06×10^8	8.61×10^7		1.82×10^7	1.74×10^6
11	1.28×10^8	1.45×10^8	1.30×10^8		1.68×10^7	2.42×10^6
12	1.70×10^8	1.81×10^8	1.75×10^8		1.14×10^7	5.81×10^6
13	2.00×10^8	2.06×10^8	2.05×10^8		5.41×10^6	5.28×10^6
14	2.08×10^8	2.12×10^8	2.09×10^8		3.98×10^6	6.81×10^5
15	1.92×10^8	2.01×10^8	1.87×10^8		9.36×10^6	4.38×10^6
16	1.58×10^8	1.77×10^8	1.52×10^8		1.86×10^7	6.65×10^6
17	1.20×10^8	1.47×10^8	1.14×10^8		2.68×10^7	5.95×10^6
18	8.64×10^7	1.18×10^8	8.26×10^7		3.13×10^7	3.86×10^6
19	5.98×10^7	9.16×10^7	5.80×10^7		3.18×10^7	1.74×10^6
20	4.05×10^7	7.01×10^7	4.03×10^7		2.96×10^7	1.97×10^5
21	2.71×10^7	5.32×10^7	2.79×10^7		2.60×10^7	7.11×10^5
22	1.81×10^7	4.02×10^7	1.93×10^7		2.20×10^7	1.14×10^6
23	1.21×10^7	3.04×10^7	1.34×10^7		1.82×10^7	1.26×10^6
24	8.17×10^6	2.30×10^7	9.38×10^6		1.48×10^7	1.21×10^6
25	5.53×10^6	1.75×10^7	6.61×10^6		1.19×10^7	1.08×10^6
26	3.77×10^6	1.33×10^7	4.70×10^6		9.57×10^6	9.31×10^5
27	2.59×10^6	1.02×10^7	3.37×10^6		7.65×10^6	7.78×10^5
28	1.79×10^6	7.90×10^6	2.43×10^6		6.11×10^6	6.39×10^5
29				FOM =	0.178	0.040

Cell E29 = SUM(E3:E28)/SUM(B3:B28)

Cell F29 = SUM(F3:F28)/SUM(C3:C28).

The FOM for the value of the parameter $E = 1.1$ eV is equal to $0.040 = 4\%$, almost four times smaller than the FOM $= 0.178 = 17.8\%$ for the case $E = 0.9$ eV.

The frequency factor s can be calculated by using the value of $E = 1.1$ eV and the temperature of maximum TL intensity $T_M = 380$ K in equation (1.9) for second-order kinetics ($b = 2$):

$$s = \frac{\beta E}{kT_M^2 \left(1 + \dfrac{2kT_M}{E}\right)} \exp\left(\frac{E}{kT_M}\right)$$

TABLE 2.16. Summary of the results of various analysis methods for
second-order data

	E (eV)	s or s'	Comments below
Initial rise method	0.969 ± 0.006		[1, 4]
Chen's τ-method	0.968		[2, 4]
Chen's δ-method	1.04 ± 0.11		[2, 4]
Chen's ω-method	1.011		[2, 4]
Whole glow-peak method	0.954 ± 0.004	$s' = 20.697$	[3]
Fitting method using Kitis et al. second-order equation (equation (2.12))	1.1 ± 0.1	$s = 1.38 \times 10^{12}$ s^{-1}	[4]

$$= \frac{1(1)}{(8.617 \times 10^{-5})(380)^2 \left(1 + \frac{2(8.617 \times 10^{-5})380}{1} \right)}$$

$$\times \exp \left(\frac{1}{(8.617 \times 10^{-5})380} \right) = 1.38 \times 10^{12} \text{ s}^{-1}. \qquad (2.13)$$

The resolution of the Kitis et al fitting method can be refined by repeating this process of calculating the FOM for different values of E spaced much closer together (e.g., $E = 1.01, 1.00, 0.99$, etc.) and then attempt to minimize the value of the FOM.

Finally, we summarize in Table 2.16 the results of the various methods for analyzing the given second-order experimental data.

Comments on the Results of Exercise 2.4

1. The value of E obtained from the IR method is independent of the kinetics of the TL glow peak.

   As in the case of first-order kinetics, the presence of thermal quenching affects the value of E obtained in the IR method.

   A possible correction method for the value of E when thermal quenching is present is given in Chapter 5.

   It is best to use the IR and peak shape methods with samples irradiated at low doses [3].

2. The value of E obtained with peak shape methods can be influenced by the presence of smaller satellite peaks.

3. The whole glow-curve method yields information on both E and the preexponential factor s. Because TL cannot yield a value for the absolute concentration n_0, the quantities $s' = s/N$ and n_0 appearing in equation (1.6) represent *empirical fitting parameters* for second-order glow peaks. The value of n_0 can be obtained from the area under the glow curve.

   By using the values of E, $s' = s/N$, and n_0 obtained from the whole glow-peak method, it is possible to compare directly the experimental data with the TL intensity obtained using equation (1.6), as was shown in this exercise.

4. The pre-exponential factor s in these methods can be calculated from the value of T_M, E, and β by using equation (2.13). The estimated uncertainties $\Delta s/s$ from equation (2.13) can be very large (50–100%), even when the uncertainty $\Delta E/E$ is very small.

(e) Can we conclude that this TL peak follows second-order kinetics?

Chen et al [4] have provided a list of criteria that should be checked before claiming that a certain TL glow peak is of second-order.

Unfortunately, the TL literature contains many publications claiming a certain kinetic order for TL glow curves, based solely on peak shape analysis of a single glow curve.

The criteria for second-order kinetics were listed by Chen et al. [4] as follows:

I. *Peak shape*: Second-order peaks exhibit $\mu = 0.52$.

II. *Peak shift*: In most non-first-order TL glow peaks, the location of maximum TL intensity shifts toward higher temperatures for lower trap filling (smaller doses). One must be aware that the observed maximum shift in the experimental data can also be due to the presence of smaller satellite peaks.

III. *Superlinearity effects*: Second-order peaks may exhibit slight superlinearity effects at low doses.

IV. *$I_M - T_M$ dependence*: In second-order peaks, a graph of $\ln\left[I_M\left(\dfrac{T_M^2}{\beta}\right)^2\right]$ versus $1/kT_M$ will yield a straight line of slope E (equation (1.29)).

V. *Isothermal decay results*: These can provide valuable independent information about the kinetics of the TL process involved at different temperatures. As discussed in Chapter 1, different kinetic orders correspond to different mathematical behaviors for the isothermal decay laws.

For second-order isothermal decay, a graph of $(I_t/I_0)^{-1/2}$ versus time should yield a straight line of slope E.

The TL-like presentation of isothermal decay data can provide useful information about the kinetics and the kinetic parameters. A numerical example using this type of presentation is given in Chapter 5.

Exercise 2.5: Isothermal Method for Second-Order Kinetics

Even though this exercise refers specifically to isothermal data following second-order kinetics, the exact same method of analysis can be used for general-order kinetics data.

A TL material is irradiated with a certain dose D, and the sample is heated rapidly to a temperature of 60°C. The temperature is then kept constant while the emitted light is measured as a function of time t. The experiment is then repeated with the same dose D and for two additional temperatures of 70°C and 80°C. The following isothermal decay data in Table 2.17 is obtained for three different temperatures T = 60°C, 70°C, and 80°C.

TABLE 2.17. Isothermal decay data for second-order kinetics

t(s)	TL, $T = 60°$C	TL, $T = 70°$C	TL, $T = 80°$C
0	6.40×10^6	1.40×10^7	2.30×10^7
100	5.57×10^6	1.02×10^7	1.25×10^7
200	4.92×10^6	7.84×10^6	8.10×10^6
300	4.20×10^6	6.18×10^6	5.45×10^6
400	3.92×10^6	4.90×10^6	3.99×10^6
500	3.54×10^6	4.13×10^6	2.90×10^6
600	3.10×10^6	3.47×10^6	2.39×10^6
700	2.91×10^6	2.95×10^6	1.94×10^6
800	2.70×10^6	2.56×10^6	1.80×10^6
900	2.44×10^6	2.21×10^6	1.34×10^6

(a) Show that these data are consistent with the assumption that this TL peak follows second-order kinetics.

(b) Find the kinetic parameters E and s.

Solution

(a) The graphs in Figure 2.20 show the given data for the three temperatures $T = 60°$C, $70°$C, and $80°$C.

We can rule out the possibility of first-order kinetics by graphing ln(TL) versus time as shown in Figure 2.21. The graphs obtained are nonlinear, indicating that the data do not conform to first-order kinetics.

As discussed in Chapter 1, the isothermal decay curves for TL peaks following general-order kinetics with a kinetic-order parameter b are given by

$$\left(\frac{I_t}{I_0}\right)^{\frac{1-b}{b}} = 1 + s'n_0^{b-1}(b-1)t \exp\left(-\frac{E}{kT}\right), \qquad (1.38)$$

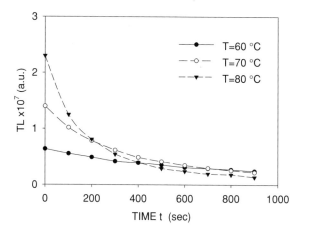

FIGURE 2.20. The isothermal data for three different temperatures and for second-order kinetics.

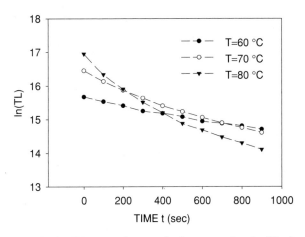

FIGURE 2.21. The ln(TL) versus time graphs for a second-order TL glow curve.

where

I_0 = initial TL intensity,
I_t = the TL intensity at time t,
$s' = s/N$ = effective frequency factor,
E = activation energy,
n_0 = initial trapped charged population,
T = temperature of isothermal decay.

This equation indicates that a plot of the quantity $(I_t/I_0)^{(1-b)/b}$ versus time t should be a straight line when a suitable value of b is found. After the determination of value of b, we will graph $(I_t/I_0)^{(1-b)/b}$ versus time t for the three different decay temperatures, and obtain a set of straight lines of slope m given by

$$m = s'n_0^{b-1}(b-1)\exp\left(-\frac{E}{kT}\right). \tag{1.39}$$

The activation energy E and the effective frequency factor $s'' = s'n_0^{b-1}$ will be determined from the slope and intercept of the plot of $\ln(m)$ versus $1/kT$.

Table 2.18 shows the calculation of the quantities $(I_t/I_0)^{(1-b)/b}$ for the isothermal decay data at $T = 70°C$, and for four different values of the kinetic-order parameter $b = 1.6$, 1.8, 2.0, and 2.2. The graph in Figure 2.22 shows these quantities as a function of time t.

It can be seen that all four graphs yield satisfactory linear fits, with the following regression coefficients:

$$b = 1.6: \quad R = 0.9986$$
$$b = 1.8: \quad R = 0.9995$$
$$b = 2.0: \quad R = 0.9999$$
$$b = 2.2: \quad R = 0.9998.$$

TABLE 2.18. Calculation of the quantities $(I_t/I_0)^{(1-b)/b}$ for the isothermal decay data

t(s)	$T = 70°C$	$(I/I_0)^{(1-b)/b}$ $b = 2.0$	$(I/I_0)^{(1-b)/b}$ $b = 1.8$	$(I/I_0)^{(1-b)/b}$ $b = 1.6$	$(I/I_0)^{(1-b)/b}$ $b = 2.2$
0	1.40×10^7	1.000	1.000	1.000	1.000
100	1.02×10^7	1.172	1.151	1.126	1.189
200	7.84×10^6	1.337	1.294	1.243	1.372
300	6.18×10^6	1.505	1.438	1.359	1.562
400	4.90×10^6	1.690	1.595	1.482	1.773
500	4.13×10^6	1.842	1.721	1.581	1.947
600	3.47×10^6	2.010	1.860	1.688	2.142
700	2.95×10^6	2.178	1.998	1.793	2.338
800	2.56×10^6	2.338	2.128	1.891	2.526
900	2.21×10^6	2.515	2.270	1.997	2.735

This example illustrates one of the possible difficulties with isothermal decay data: It may be difficult to obtain an exact estimate of the best linear fit, because small differences may occur between the graphs for various values of b. The above values of R indicate that the graph corresponding to $b = 2.0$ represents the best linear fit, and therefore the given TL data are consistent with second-order kinetics.

The above type of analysis must be carried out for all available isothermal decay data. Once the kinetic order b is ascertained by the above type of analysis, we graph $(I_t/I_0)^{(1-b)/b} = (I_t/I_0)^{-1/2}$ versus time t, and find the slopes of the resulting linear graphs.

These graphs are shown in Figure 2.23 for the available isothermal decay data at $T = 60°C$, $70°C$, and $80°C$. We next find the regression lines through each of the graphs in Figure 2.23.

(b) We now tabulate in Table 2.19 the slopes of these linear graphs and calculate the natural logarithm of the slopes, ln(slope). Finally, we graph in Figure 2.24 the

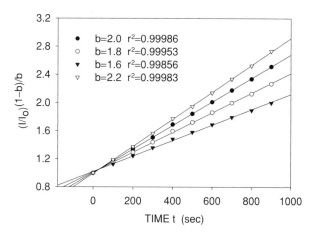

FIGURE 2.22. Graphs calculated for several values of kinetic order b.

TABLE 2.19. The slopes of linear isothermal graphs
and their natural logarithms ln(slope)

$T(°C)$	slope (s^{-1})	$1/kT$ (eV^{-1})	ln (Slope)
60	6.9580×10^{-4}	34.8498	−7.2704
70	1.7310×10^{-4}	33.8337	−6.3591
80	3.4560×10^{-4}	32.8753	−5.6676

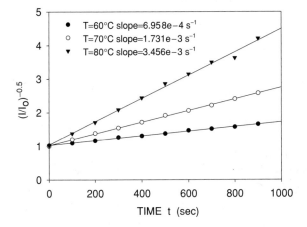

FIGURE 2.23. Calculated graphs for kinetic order $b = 2.0$ and for second-order TL data.

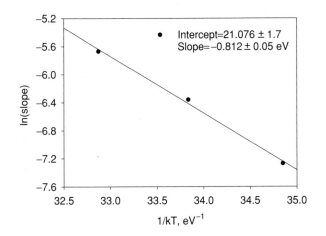

FIGURE 2.24. The ln(slope) versus $1/kT$ graph to determine E for second-order TL data.

ln(slope) versus $1/kT$, where T = temperature (K) at which the isothermal decay curves were measured.

The slope of the regression line gives the activation energy E:

$$E = 0.812 \pm 0.050 \, \text{eV}.$$

And the frequency factor $s'' = s'n_0^{b-1}$ can be found from the intercept of the regression line:

$$\text{Intercept} = \ln((b-1)\,s'') = 21.1 \pm 1.7.$$

Therefore, by substituting in our case b = 2,

$$s'' = \exp(21.1) = 1.46 \times 10^9 \, s^{-1}.$$

The errors $\Delta s''$ can be calculated from the uncertainties in the intercept of the regression line as follows:

$$\Delta \, (\text{intercept}) = \frac{\partial(\ln s'')}{\partial s''}\Delta s'' = \frac{\Delta s''}{s''} = 1.7.$$

This gives a typical large error for the value of the effective frequency factor $s'' = s'n_0^{b-1}$.

Once again, it is noted that the parameter $s'' = s'n_0^{b-1}$ cannot yield any additional information on the kinetics of the TL process, but rather represents an empirical fitting parameter for equation (1.38).

Exercise 2.6: Analysis of a General-Order TL Peak

You are given the experimental data in Table 2.20 and Figure 2.25 for a TL glow curve (TL versus temperature T), and the known heating rate $\beta = 1\text{K s}^{-1}$.

(a) Apply the IR method to find the activation energy E. The value for E obtained using the IR method is assumed to be independent of the order of kinetics.

TABLE 2.20. Experimental data for general-order kinetics TL peak

$T(^\circ C)$	$TL_\text{experimental}$	$T(^\circ C)$	$TL_\text{experimental}$
0	8.38×10^5	130	3.96×10^7
10	2.06×10^6	140	1.96×10^7
20	4.74×10^6	150	9.19×10^6
30	1.03×10^7	160	4.22×10^6
40	2.09×10^7	170	1.93×10^6
50	3.98×10^7		
60	6.97×10^7		
70	1.09×10^8		
80	1.49×10^8		
90	1.70×10^8		
100	1.57×10^8		
110	1.18×10^8		
120	7.32×10^7		

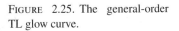

FIGURE 2.25. The general-order
TL glow curve.

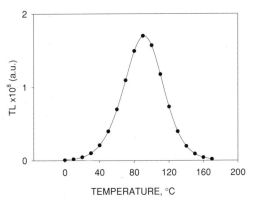

(b) Apply Chen's peak shape equations to find the activation energy E, using the shape parameters τ, δ, and ω. By assuming that the experimental error in the quantities τ, δ, and ω is $\Delta T = 2$ K estimate the error $\Delta\mu$ in the value of the geometrical shape factor μ.

Show that the values of μ and $\Delta\mu$ are consistent with the assumption that the TL glow curve obeys general-order kinetics.

(c) By using the experimental data, apply the whole glow-peak method to find E, s, and the order of kinetics b. Graph $\ln(I/n^b)$ versus $1/T$ for various values of b and find the correct value of b that gives a linear graph.

From the slope and intercept of the graph $\ln(I/n^b)$ versus $1/T$, calculate the kinetic parameters.

Verify that the given TL glow curve corresponds to general-order kinetics.

(d) Using the experimental values of I_M (maximum TL intensity) and T_M (temperature of maximum intensity), do a curve fitting to the given numerical data. Use the following analytical equation developed by Kitis et al [2] for general-order kinetic peaks. The expression relies on two experimentally measured quantities I_M and T_M:

$$I(T) = I_M b^{b-1} \exp\left(\frac{E}{kT} \cdot \frac{T - T_M}{T_M}\right)$$

$$\times \left[(b-1)\frac{T^2}{T_M^2}\left(1 - \frac{2kT}{E}\right)\exp\left(\frac{E}{kT}\cdot\frac{T - T_M}{T_M}\right)\right.$$

$$\left. + 1 + (b-1)\frac{2kT_M}{E}\right]^{-\frac{b}{b-1}}. \tag{2.14}$$

The activation parameter E can be treated as an adjustable parameter. Graph both the experimental data and the calculated general order TL glow curve on the same graph and compare them. Calculate the FOM for the TL glow curve.

(e) Can it be concluded from the above analysis that this material will always follow general-order kinetics?

TABLE 2.21. The values of $1/kT$ and of the natural logarithm of the TL data, $\ln(TL)$

$T(°C)$	$TL_{experimental}$	$1/kT(eV^{-1})$	$\ln(TL)$	$T(°C)$	$TL_{experimental}$	$1/kT(eV^{-1})$	$\ln(TL)$
0	8.38×10^5	42.51	13.64	130	3.96×10^7	28.80	17.49
10	2.06×10^6	41.01	14.54	140	1.96×10^7	28.10	16.79
20	4.74×10^6	39.61	15.37	150	9.19×10^6	27.43	16.03
30	1.03×10^7	38.30	16.15	160	4.22×10^6	26.80	15.26
40	2.09×10^7	37.08	16.86	170	1.93×10^6	26.20	14.47
50	3.98×10^7	35.93	17.50				
60	6.97×10^7	34.85	18.06				
70	1.09×10^8	33.83	18.51				
80	1.49×10^8	32.88	18.82				
90	1.70×10^8	31.97	18.95				
100	1.57×10^8	31.11	18.87				
110	1.18×10^8	30.30	18.58				
120	7.32×10^7	29.53	18.11				

Solution

(a) *The IR method.* We calculate in Table 2.21 the values of $1/kT$ (T = temperature (K), k = Boltzman constant) and the values of the natural logarithm of the TL data, $\ln(TL)$.

We next graph the $\ln(TL)$ versus $1/kT$ data and find a regression line through the first 7 data points, as shown in Figure 2.26.

The slope of the regression line gives the activation energy E as

$$E = 0.580 \pm 0.006 \text{ eV}, \quad \text{with } R^2 = 0.9996.$$

(b) *Chen's peak shape equations.* From the given experimental data for a TL glow peak, we can estimate the three temperatures required for Chen's peak shape equations:

$$T_1 = 64°C = 337 \text{ K}, \quad T_2 = 117°C = 390 \text{ K}, \quad T_M = 91°C = 364 \text{ K},$$

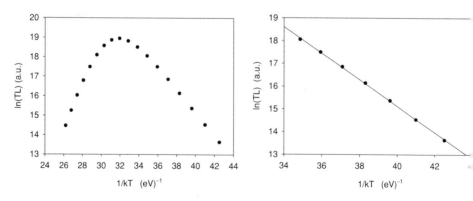

FIGURE 2.26. The IR analysis for general-order kinetics data.

where

T_M = peak temperature at the maximum TL intensity,

T_1, T_2 = temperatures on either side of T_M, corresponding to the half-maximum intensity.

We next calculate the quantities μ, τ, δ, and ω:

$$\tau = T_M - T_1 = 27\,\text{K}, \quad \delta = T_2 - T_M = 26\,\text{K},$$
$$\omega = T_2 - T_1 = 53\,\text{K}, \quad \mu = \delta/\omega = 26/53 = 0.490.$$

The calculated value of the geometrical shape factor $\mu = \delta/\omega = 0.490$ corresponds to a value of μ for a general-order TL peak with an approximate value of $b = 1.6$ (Figure 1.14).

Using the known experimental error $\Delta T = 2\,\text{K}$ for the quantities τ, δ, and ω, we can do an error analysis of the values of μ. As in the similar examples for first- and second-order kinetics data,

$$\left|\frac{\Delta\mu}{\mu}\right| = \left|\frac{\Delta\delta}{\delta}\right| + \left|\frac{\Delta\omega}{\omega}\right| = \left|\frac{2}{26}\right| + \left|\frac{2}{53}\right| = 0.077 + 0.038 = 0.115.$$

This leads to a value of $\mu + \Delta\mu = 0.490 \pm 0.056$ which is consistent with general-order kinetics, within the accuracy of the TL experiment.

In order to find the activation energy E, we apply Chen's equation for general-order kinetics:

$$E_\alpha = c_\alpha \left(\frac{kT_M^2}{\alpha}\right) - b_\alpha(2kT_M), \tag{2.15}$$

where α is τ, δ, or ω and the values of c_α and b_α are summarized below

$$c_\tau = 1.510 + 3.0(\mu - 0.42), \quad b_\tau = 1.58 + 4.2(\mu - 0.42)$$
$$c_\delta = 0.976 + 7.3(\mu - 0.42), \quad b_\delta = 0$$
$$c_\omega = 2.52 + 10.2(\mu - 0.42), \quad b_\omega = 1. \tag{2.16}$$

Using the value of τ:

$$E = \frac{1.720kT_M^2}{\tau} - 1.874(2kT_M) = 0.727 - 0.117 = 0.610\,\text{eV}.$$

Using the value of δ:

$$E = \frac{1.487kT_M^2}{\delta} = 0.653\,\text{eV}.$$

Using the value of ω:

$$E = \frac{3.234kT_M^2}{\omega} - 2kT_M = 0.671 - 0.062 = 0.609\,\text{eV}.$$

TABLE 2.22. Calculation of the quantities $\ln(I/n_b)$ and $1/kT$ for general-order data

$T(^\circ C)$	TL	$1/kT$	Area	$\ln(\text{TL}/n^{1.2})$	$\ln(\text{TL}/n^{1.4})$	$\ln(\text{TL}/n^{1.5})$	$\ln(\text{TL}/n^{1.6})$
0	8.38×10^5	42.51	9.99×10^8	-13.99	-18.60	-20.90	-23.20
10	2.06×10^6	41.01	9.98×10^8	-13.09	-17.70	-20.00	-22.30
20	4.74×10^6	39.61	9.96×10^8	-12.25	-16.86	-19.16	-21.46
30	1.03×10^7	38.30	9.91×10^8	-11.47	-16.08	-18.38	-20.68
40	2.09×10^7	37.08	9.81×10^8	-10.75	-15.35	-17.65	-19.95
50	3.98×10^7	35.93	9.60×10^8	-10.08	-14.68	-16.98	-19.28
60	6.97×10^7	34.85	9.20×10^8	-9.47	-14.06	-16.35	-18.65
70	1.09×10^8	33.83	8.50×10^8	-8.93	-13.50	-15.78	-18.07
80	1.49×10^8	32.88	7.41×10^8	-8.45	-13.00	-15.27	-17.54
90	1.70×10^8	31.97	5.92×10^8	-8.05	-12.55	-14.80	-17.05
100	1.57×10^8	31.11	4.22×10^8	-7.72	-12.16	-14.37	-16.59
110	1.18×10^8	30.30	2.65×10^8	-7.46	-11.80	-13.97	-16.14
120	7.32×10^7	29.53	1.48×10^8	-7.23	-11.45	-13.56	-15.67
130	3.96×10^7	28.80	7.45×10^7	-7.02	-11.11	-13.15	-15.19
140	1.96×10^7	28.10	3.49×10^7	-6.82	-10.75	-12.72	-14.68
150	9.19×10^6	27.43	1.53×10^7	-6.58	-10.35	-12.24	-14.12
160	4.22×10^6	26.80	6.15×10^6	-6.27	-9.85	-11.65	-13.44
170	1.93×10^6	26.20					

In order to find the error ΔE in the activation energy E, we take the logarithmic derivative of the equation $E = \dfrac{1.487kT_M^2}{\delta}$:

$$\left|\frac{\Delta E}{E}\right| = 2\left|\frac{\Delta T_M}{T_M}\right| + \left|\frac{\Delta \delta}{\delta}\right| = 2\left|\frac{2}{364}\right| + \left|\frac{2}{26}\right| = 0.011 + 0.077 = 0.088.$$

This gives an error ΔE of the order of 8.8% or $\Delta E = 0.088E = 0.082(0.653) = 0.060$ eV.

(c) *The whole glow-peak method.* We graph $\ln(I/n^b)$ versus $1/T$ for various values of b between 1.2 and 1.6, and find the correct value of b that gives a linear graph. As in the case of first-order kinetics, $n(T)$ is the area under the glow peak and it is calculated starting at a temperature T, up to the maximum temperature available in the glow curve. In the data shown in Table 2.22, the maximum available temperature is 170°C.

By following the same procedure as in the case of first-order kinetics, we set up an Excel spreadsheet to calculate the quantities $\ln(I/n^b)$ and $1/kT$ as shown in Table 2.22, for several values of the kinetic order $b = 1.2, 1.4, 1.5$ and 1.6.

Finally, graphs of $\ln(\text{TL/Area}^b)$ versus $1/kT$ are drawn in Figure 2.27 for several values of the kinetic order $b = 1.2, 1.4, 1.5$, and 1.6.

The graphs corresponding to $b = 1.5$ and 1.6 best approximate straight lines. The $b = 1.5$ graph has the highest value of R^2 and therefore gives the best fit. A regression line is fitted to the data corresponding to $b = 1.5$ in Figure 2.28, to obtain the best slope and the best intercept:

$$\text{Best intercept} = 3.345 \pm 0.17,$$
$$\text{Best slope } E = -0.568 \pm 0.005 \text{ eV}.$$

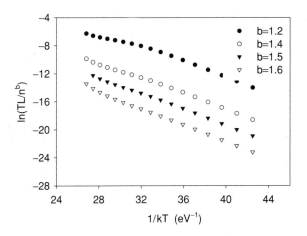

FIGURE 2.27. Graphs for several values of kinetic order b.

It is noted that this value of $E = 0.568 \pm 0.005$ eV is in good agreement with the value of $E = 0.580 \pm 0.006$ eV obtained from the IR method.

Within the accuracy of the given experimental data and within the framework of the whole glow-peak method of analysis, we can conclude that the given TL glow peak follows general-order kinetics described by $b = 1.5$.

The value of s' can be calculated from the best-fitting regression line shown in Figure 2.28:

$$s' = \beta e^{(\text{intercept})} = 1 e^{(3.345)} = 28.36.$$

The whole glow-peak method yields information about both the activation energy E and the effective frequency factors s' and $s'' = s' n_0^{b-1}$ appearing in

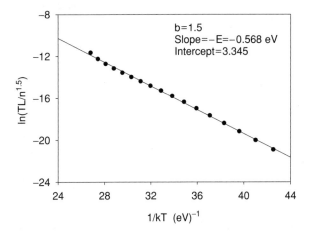

FIGURE 2.28. The value of $b = 1.5$ gives the best fit for the whole glow-curve analysis. The parameters E and s' can be calculated from this graph.

equation (1.7):

$$I(T) = s'' n_0 \exp\left(-\frac{E}{kT}\right) \left[1 + \frac{s''(b-1)}{\beta} \int_{T_0}^{T} \exp\left(-\frac{E}{kT'}\right) dT'\right]^{-\frac{b}{b-1}}. \quad (1.7)$$

Because TL data analysis of a glow curve cannot yield a value for the *absolute* concentration n_0 of traps in the material, the quantities s'', b, and n_0 appearing in equation (1.7) represent *three empirical fitting parameters* for general-order glow peaks.

The value of n_0 can be estimated from the area under the glow curve as in Exercise 2.4, by summing the TL intensities multiplied by the temperature interval ΔT between TL measurements, and by dividing with the heating rate β:

$$n_0 \approx \frac{1}{\beta} \int_{T_0}^{T_f} I \, dT = \frac{1}{\beta} \sum I(T)\Delta T = \frac{1}{1} \sum I(T)(10\,\text{K}) = 9.99 \times 10^9.$$

By using the values of $E = 0.56$ eV, $s' = 28.36$, $n_0 = \text{Area} = 9.99 \times 10^9$, and $\beta = 1$ K s^{-1}, it is possible to calculate the TL intensity using equation (1.7), and to compare this result directly with the given experimental data.

The result is shown in Figure 2.27, where the calculated $I(T)$ from equation (1.7) is compared with the original experimental data. Figure 2.29 shows that the calculated parameters E, s', and b and n_0 from the whole glow-peak method, as well as the general order equation (1.7), describe the given experimental data in a reasonably accurate manner.

The observed discrepancies between experiment and calculation in Figure 2.29 are due to the several approximations involved in applying the whole glow-peak method.

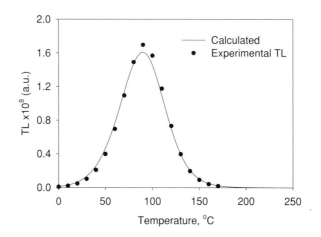

FIGURE 2.29. Comparison of calculated TL intensity using equation (1.7) (solid line), and original experimental data (individual data points). The parameters used in equation (1.7) were obtained from the whole glow-peak method.

(d) *Glow-curve fitting using the Kitis et al equation.* We use the following analytical equation developed by Kitis et al [2] for TL peaks following general-order kinetics. The expression relies on two experimentally measured quantities I_M (the maximum TL intensity) and T_M (the temperature corresponding to the maximum TL intensity):

$$I(T) = I_M b^{b-1} \exp\left(\frac{E}{kT} \cdot \frac{T - T_M}{T_M}\right)$$

$$\times \left[(b-1)\frac{T^2}{T_M^2} \cdot \left(1 - \frac{2kT}{E}\right) \exp\left(\frac{E}{kT} \cdot \frac{T - T_M}{T_M}\right) \right.$$

$$\left. + 1 + (b-1)\frac{2kT_M}{E} \right]^{-\frac{b}{b-1}}. \tag{2.17}$$

For the given experimental data, $T_M = 364$ K and $I_M = 1.70 \times 10^8$. By treating the activation parameter E as an adjustable parameter, we calculate several graphs with values of $E = 0.4, 0.5, 0.6$, and 0.7 eV. The calculations can be set up easily in a spreadsheet as shown in Table 2.23.

Columns A–C contain the experimental data points for the TL glow curve, whereas columns D–G contain the calculated data points using equation (2.17) for general-order kinetics $b = 1.5$ and for four values of the energy parameter $E(E = 0.4, 0.5, 0.6,$ and 0.7 eV).

The following equation is used to calculate the values of the fitted data in column D, using equation (2.17) for general-order kinetics:

Cell D3 = ($I$8^($I$8/($I$8-1)))*$I$6*EXP($I$2/(0.00008617*B3)
  *((B3-$I$5)/$I$5))*((B3*B3)/($I$5*$I$5)
  *(1-2*0.00008617*B3/$I$2)*EXP($I$2/(0.00008617*B3)
  *((B3-$I$5)/$I$5))*($I$8-1)+1+($I$8-1)
  *(2*0.00008617*$I$5/$I$2))^-($I$8/($I$8-1)).

This expression refers to cell B3 which contains the absolute temperature $T(K)$. Also, note that cell I2 in the spreadsheet contains the value of the energy parameter $E = 0.4$ eV, cell I5 contains the value of the experimental parameter $T_M = 364$ K, and cell I6 contains the value of the experimental maximum height parameter $I_M = 1.70 \times 10^8$. The parameter $b = 1.5$ is contained in cell I8. The above spreadsheet expression refers to the values contained in these cells by using the Excel expressions $I$2, $I$5, $I$6, and $I$8, correspondingly.

The user controls the value of the parameter E by changing the value in cell I2, and the whole spreadsheet calculation is automatically updated.

The graphs calculated for $E = 0.4, 0.5, 0.6,$ and 0.7 eV are shown in Figure 2.30.

It can be seen in Figure 2.30 that when the value of E is too low (graph corresponding to $E = 0.4$, and 0.5 eV), the calculated TL points lie well above the experimental data. This is also evident by inspection of the calculated columns

TABLE 2.23. Calculations for several values of energy E

	A	B	C	D	E	F	G	H	I	J
	$T(°C)$	$T(K)$	TL$_{\text{experimental}}$	$I(T)$, $E = 0.4$ eV	$I(T) = 0.5$ eV	$I(T) = 0.6$ eV	$I(T) = 0.7$ eV			
1										
2								$E =$	0.4	eV
3	0	273	8.38×10^5	6.44×10^6	2.34×10^6	8.35×10^5	2.95×10^5			
4	10	283	2.06×10^6	1.16×10^7	4.94×10^6	2.05×10^6	8.44×10^5			
5	20	293	4.74×10^6	2.00×10^7	9.85×10^6	4.73×10^6	2.24×10^6	$T_M =$	364	K
6	30	303	1.03×10^7	3.29×10^7	1.86×10^7	1.03×10^7	5.56×10^6	$I_M =$	1.70×10^8	
7	40	313	2.09×10^7	5.12×10^7	3.32×10^7	2.09×10^7	1.29×10^7			
8	50	323	3.98×10^7	7.55×10^7	5.55×10^7	3.97×10^7	2.78×10^7	$b =$	1.5	
9	60	333	6.97×10^7	1.04×10^8	8.61×10^7	6.94×10^7	5.49×10^7			
10	70	343	1.09×10^8	1.34×10^8	1.22×10^8	1.09×10^8	9.64×10^7			
11	80	353	1.49×10^8	1.58×10^8	1.54×10^8	1.49×10^8	1.43×10^8			
12	90	363	1.70×10^8	1.69×10^8	1.69×10^8	1.69×10^8	1.69×10^8			
13	100	373	1.57×10^8	1.65×10^8	1.61×10^8	1.57×10^8	1.52×10^8			
14	110	383	1.18×10^8	1.46×10^8	1.32×10^8	1.18×10^8	1.03×10^8			
15	120	393	7.32×10^7	1.18×10^8	9.47×10^7	7.35×10^7	5.52×10^7			
16	130	403	3.96×10^7	8.79×10^7	6.08×10^7	3.99×10^7	2.50×10^7			
17	140	413	1.96×10^7	6.17×10^7	3.61×10^7	1.98×10^7	1.03×10^7			
18	150	423	9.19×10^6	4.12×10^7	2.03×10^7	9.32×10^6	4.10×10^6			
19	160	433	4.22×10^6	2.67×10^7	1.10×10^7	4.30×10^6	1.61×10^6			
20	170	443	1.93×10^6	1.69×10^7	5.93×10^6	1.97×10^6	6.38×10^5			

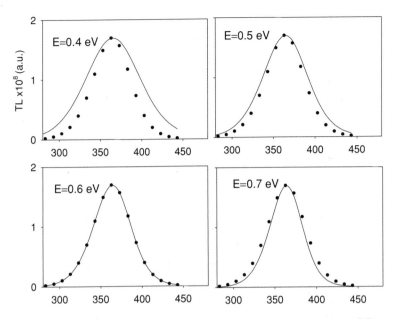

FIGURE 2.30. The experimental data and fitted curves for several values of E.

$I(T)$ in Table 2.23. On the other hand, when the value of E is too high (graph corresponding to $E = 0.7$ eV), the calculated TL points lie well below the experimental data.

Table 2.24 shows the FOM calculation as applied to the previous data. Column A contains the experimental data points and columns B–E contain the calculated data points using equation (2.17) for general-order kinetics and for four values of the energy parameter E (E $= 0.4, 0.5, 0.6$, and 0.7 eV).

Columns F–I contain the calculation of the quantity $|\text{TL}_{\text{experimental}} - \text{TL}_{\text{calculated}}|$, and the cells in the last row contain the calculated values of the FOM.

The FOM for the value of the parameter $E = 0.60$ eV is equal to $0.003 = 0.3\%$, almost 100 times smaller than the FOM $= 0.300 = 30.0\%$ for the case $E = 0.4$ eV.

The frequency factor s can be calculated by using the value of $E = 0.60$ eV and the temperature of maximum TL intensity $T_M = 364$ K in equation (1.10), which is applicable for general-order kinetics:

$$s = \frac{\beta E}{kT_M^2 \left(1 + \dfrac{2kT_M(b-1)}{E}\right)} \exp\left(\frac{E}{kT_M}\right)$$

$$s = \frac{(1)0.60}{(8.617 \times 10^{-5})(364)^2 \left(1 + \dfrac{2(8.617 \times 10^{-5})364(1.5-1)}{0.60}\right)}$$

$$\times \exp\left(\frac{0.60}{(8.617 \times 10^{-5})364}\right) = 1.05 \times 10^{12}\, s^{-1}. \qquad (2.18)$$

TABLE 2.24. The FOM calculation for general-order kinetics TL peak

A	B	C	D	E	F	G	H	I
$TL_{experimental}$	$I(T)$, $E = 0.4$ eV	$I(T)$ $E = 0.5$ eV	$I(T)$ $E = 0.6$ eV	$I(T)$ $E = 0.7$ eV	$\|TL_{experimental} - I(T)\|$ $E = 0.4$ eV	$\|TL_{experimental} - I(T)\|$ $E = 0.5$ eV	$\|TL_{experimental} - I(T)\|$ $E = 0.6$ eV	$\|TL_{experimental} - I(T)\|$ $E = 0.7$ eV
8.38×10^5	6.44×10^6	2.34×10^6	8.35×10^5	2.95×10^5	5.61×10^6	1.50×10^6	2.61×10^3	5.43×10^5
2.06×10^6	1.16×10^7	4.94×10^6	2.05×10^6	8.44×10^5	9.58×10^6	2.88×10^6	6.48×10^3	1.22×10^6
4.74×10^6	2.00×10^7	9.85×10^6	4.73×10^6	2.24×10^6	1.53×10^7	5.11×10^6	1.47×10^4	2.50×10^6
1.03×10^7	3.29×10^7	1.86×10^7	1.03×10^7	5.56×10^6	2.26×10^7	8.32×10^6	3.19×10^4	4.72×10^6
2.09×10^7	5.12×10^7	3.32×10^7	2.09×10^7	1.29×10^7	3.03×10^7	1.22×10^7	6.71×10^4	8.05×10^6
3.98×10^7	7.55×10^7	5.55×10^7	3.97×10^7	2.78×10^7	3.57×10^7	1.57×10^7	1.30×10^5	1.20×10^7
6.97×10^7	1.04×10^8	8.61×10^7	6.94×10^7	5.49×10^7	3.46×10^7	1.65×10^7	2.28×10^5	1.47×10^7
1.09×10^8	1.34×10^8	1.22×10^8	1.09×10^8	9.64×10^7	2.43×10^7	1.23×10^7	3.46×10^5	1.30×10^7
1.49×10^8	1.58×10^8	1.54×10^8	1.49×10^8	1.43×10^8	8.61×10^6	4.49×10^6	4.15×10^5	5.93×10^6
1.70×10^8	1.69×10^8	1.69×10^8	1.69×10^8	1.69×10^8	3.03×10^5	3.02×10^5	3.24×10^5	3.64×10^5
1.57×10^8	1.65×10^8	1.61×10^8	1.57×10^8	1.52×10^8	8.04×10^6	4.23×10^6	5.75×10^4	4.84×10^6
1.18×10^8	1.46×10^8	1.32×10^8	1.18×10^8	1.03×10^8	2.82×10^7	1.46×10^7	2.22×10^5	1.44×10^7
7.32×10^7	1.18×10^8	9.47×10^7	7.35×10^7	5.52×10^7	4.45×10^7	2.15×10^7	3.46×10^5	1.80×10^7
3.96×10^7	8.79×10^7	6.08×10^7	3.99×10^7	2.50×10^7	4.83×10^7	2.12×10^7	3.14×10^5	1.46×10^7
1.96×10^7	6.17×10^7	3.61×10^7	1.98×10^7	1.03×10^7	4.21×10^7	1.65×10^7	2.20×10^5	9.23×10^6
9.19×10^6	4.12×10^7	2.03×10^7	9.32×10^6	4.10×10^6	3.21×10^7	1.11×10^7	1.34×10^5	5.09×10^6
4.22×10^6	2.67×10^7	1.10×10^7	4.30×10^6	1.61×10^6	2.25×10^7	6.82×10^6	7.59×10^4	2.61×10^6
1.93×10^6	1.69×10^7	5.93×10^6	1.97×10^6	6.38×10^5	1.50×10^7	4.00×10^6	4.11×10^4	1.29×10^6
				FOM $=$	0.300	0.152	0.003	0.154

Table 2.25. Summary of the results of various analysis methods for general-order data

	E (eV)	Frequency factor (s or s'')	Comments below
Initial rise method	0.580 ± 0.006		[1,4]
Chen's τ-method	0.610		[2,4]
Chen's δ-method	0.653 ± 0.06		[2,4]
Chen's ω-method	0.609		[2,4]
Whole glow-peak method	0.568 ± 0.005	$s' = 28.36$	[3]
Fitting method using Kitis et al equation	0.60 ± 0.1	$s = 1.05 \times 10^{12}\ \mathrm{s}^{-1}$	[4]

Finally, we summarize in Table 2.25 the results of the various methods for analyzing the given experimental data for general-order kinetics.

Comments

1. The value of E obtained from the IR method is independent of the kinetics of the TL glow peak.

   As in the case of first- and second-order kinetics, the presence of thermal quenching affects the value of E obtained in the IR method.

   A possible correction method for the value of E is given in Chapter 5.

   It is best to use the IR and peak shape methods with samples irradiated at low doses [3].
2. The value of E obtained with peak shape methods can be influenced by the presence of smaller satellite peaks.
3. The whole glow-curve method yields information on both E and the pre-exponential factor $s'' = n_0^{b-1} s'$. Because TL cannot yield a value for the absolute concentration n_0^{b-1}, the factor s'' and the n_0 appearing in equation (1.7) represent *two empirical fitting parameters* for general-order glow peaks. The value of n_0 can be estimated from the area under the glow curve.

   By using the values of E, s'', and n_0, it is possible to compare directly the experimental data with the TL intensity obtained using equation (1.7).
4. The pre-exponential factor s can be calculated from the values of T_M, E, and β by using equation (2.18). The estimated uncertainties $\Delta s/s$ from equation (2.18) can be very large (50–100%), even when the uncertainty $\Delta E/E$ is very small.

(e) Can it be assumed for this material that this TL peak will always follow general kinetics of order $b = 1.5$?

   In general, one cannot assume that the studied TL glow curve of this material will always follow general-order kinetics of the same order found in analyzing one set of data. The analysis should be carried out for glow peaks measured under different heating rates, various irradiation doses, powdered and bulk samples, etc.

Some of the criteria for general-order kinetics are:

I. *Peak shape*: first-order peaks have $\mu = 0.42$, second-order peaks have $\mu = 0.52$, and general-order kinetics have values in-between (see Figure 1.14).

II. *Peak shift*: In most non-first-order TL glow peaks, the location of maximum TL intensity shifts toward higher temperatures for lower trap filling. One must be aware that the observed maximum shift can be also due to the presence of smaller satellite peaks.

III. $I_M - T_M$ *dependence*: In general-order peaks a graph of $\ln \left[I_M{}^{b-1} \left(\dfrac{T_M^2}{\beta} \right)^b \right]$ versus $1/kT_M$ will yield a straight line of slope E (equation (1.28)).

IV. *Isothermal decay results*: These can provide valuable independent information about the kinetics of the TL process involved at different temperatures. General-order kinetics corresponds to decay curves described by a plot of the quantity $\left(\dfrac{I_t}{I_0} \right)^{\frac{1-b}{b}}$ versus time, which should be a straight line when a suitable value of b is found.

Exercise 2.7: Influence of the Background on the Results of the IR Method

Given the experimental data of Table 2.26, estimate the influence of the background on the activation energy obtained using the IR method.

Solution

In Exercise 2.1, it was found that the activation energies calculated using the IR method depend strongly on the number of points chosen, i.e. on the TL intensity I_{start} at the starting temperature of the IR region. From a statistical point of view, the IR region must start from the temperature at which the TL intensity is higher than the background by at least three times the standard deviation of the background signal.

TABLE 2.26. Experimental data-effect of background on IR method

T(K)	TL	T(K)	TL	T(K)	TL	T(K)	TL	T(K)	TL
293.4	5	341	6	353.8	11	357.8	25	379.8	228
297.8	3	345.8	4	354.2	25	358.2	21	382.6	316
302.6	2	350.6	7	354.6	19	360.6	25	385	400
307.4	8	351	8	355	13	363	41	387.4	538
312.2	4	351.4	9	355.4	14	365.4	51	389.8	646
317	4	351.8	10	355.8	11	367.8	46	396.6	1420
321.8	1	352.2	16	356.2	20	370.2	79	401	2152
326.6	7	352.6	15	356.6	25	352.6	110	405.8	3452
331.4	5	353	12	357	15	375	128	410.2	5127
336.2	5	353.4	16	357.4	27	377.4	197	413	6483

TABLE 2.27. Data for the two IR lines

$1/kT$	ln(TL)	ln(TL − bg)	$1/kT$	ln(TL)	ln(TL − bg)
32.58	3.00	2.69	30.95	4.85	4.81
32.54	3.22	2.99	30.75	5.28	5.26
32.51	2.71	2.28	30.56	5.43	5.51
32.47	3.30	3.08	30.33	5.76	5.74
32.43	3.22	2.99	30.14	5.99	5.98
32.40	3.04	2.76	29.96	6.29	6.28
32.18	3.22	2.99	29.77	6.47	6.46
31.97	3.71	3.58	29.26	7.26	7.25
31.76	3.93	3.82	28.94	7.67	7.67
31.55	3.83	3.71	28.60	8.15	8.15
31.35	4.37	4.30	28.29	5.84	8.54
31.15	4.70	4.65	28.10	8.78	8.78

By examining the given data, it is clear that the TL intensity at low temperatures is due to the background only. Therefore, the first 15 points, i.e. the TL in the temperature region [293–350 K] can be used to evaluate the mean value of the background. It is found by averaging these 15 points that the background is equal to bg = 5.2 ± 2.3 counts.

Once the background is evaluated and in order to show its effect on the activation energy calculation, the IR linear fit is applied twice. In the first case, the fit is performed using the TL data as given, by graphing ln(TL) versus $1/kT$, and in the second case, by subtracting the background and graphing ln(TL − bg) versus $1/kT$. The corresponding data for the two fits are shown in Table 2.27.

The resulting IR lines are shown in Figure 2.31, where one can see the different slopes obtained. The activation energy values obtained from each IR line are:

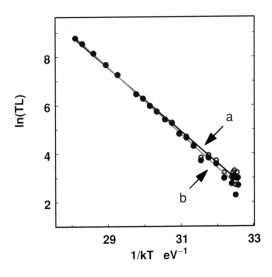

FIGURE 2.31. IR plots: (a) without background subtraction and (b) with background subtracted.

(a) Without background subtraction: $E = 1.3087 \pm 0.022$.
(b) With background subtracted: $E = 1.3780 \pm 0.024$.

It is seen that by ignoring the background, the IR method in this exercise leads to a serious underestimation of the activation energy E by 5%.

Exercise 2.8: Study of the 15% Rule of Thumb for the IR Technique

A well-known rule of thumb for the IR method is that the method holds only up to a temperature that corresponds to a TL intensity lower than 10–15% of the peak maximum intensity, I_M.

Calculate a synthetic glow peak with the trapping parameters $E = 1$ eV, $s = 10^{12}$ s^{-1}, $n_0 = 10^6$ m^{-3} and verify the applicability of the 15% rule by following these steps:

(a) Express the rule of thumb in a mathematical form by examining the two terms in the analytical expression for first-order TL glow peaks.
(b) Express the mathematical condition in (a) as a function of the percent ratio $\% I_{IR}/I_M = (I_{IR}/I_M) \times 100\%$ where I_{IR} is the maximum TL intensity of the IR region.
(c) Examine how the activation energy values obtained by the IR method depend not only on the end of the IR region (i.e. on the $\% I_{IR}/I_M$ value), but also on the starting TL intensity I_{start} of the IR region.

Solution

(a) By substituting the series approximation for the TL integral from equation (1.52) into equation (1.5) for first-order TL glow peaks, we obtain the following analytical expression:

$$I(T) = n_0 \cdot s \cdot \exp\left(-\frac{E}{kT}\right) \cdot \exp\left[-\frac{skT^2}{\beta E} \cdot \exp\left(-\frac{E}{kT}\right) \cdot \left(1 - \frac{2kT}{E}\right)\right].$$

(2.19)

This expression consists of two parts, the increasing IR part $n_0 s \exp(-E/kT)$ and the function $F2(T)$ given by

$$F2(T) = \exp\left[-\frac{skT^2}{\beta E} \cdot \exp\left(-\frac{E}{kT}\right) \cdot \left(1 - \frac{2kT}{E}\right)\right].$$

(2.20)

The IR method holds when the condition $F2(T) \sim 1$ is fulfilled.

Using a spreadsheet program, it is straight forward to evaluate a single TL glow peak with the given parameters, as well as the values of the expression $F2(T)$ at various temperatures T. The result is shown in Figure 2.32 where one can identify the part of the glow peak which must be used for the correct application of the IR method.

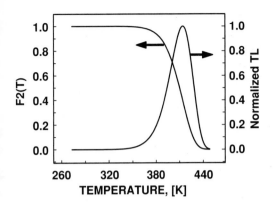

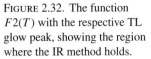

FIGURE 2.32. The function $F2(T)$ with the respective TL glow peak, showing the region where the IR method holds.

(b) The commonly used rule of thumb for the IR method says that the method holds up to the temperature at which the TL intensity is less than 10–15% of the peak maximum intensity I_M. From the data shown in Figure 2.32, one can extract the data for a plot of $F2(T)$ versus the percent ratio $\%I_{IR}/I_M$. The result is shown in Figure 2.33 from which one can conclude that the 15% rule for the IR method holds when $F2(T) > 0.95$.

(c) The IR method is applied to the synthetic glow peak by varying the starting TL intensity I_{start}. The upper limit $\%I_{IR}/I_M$ was left to vary and an IR plot was accepted if its correlation coefficient was better than 0.999. Figure 2.34 shows the resulting activation energies E as a function of the percent ratio $\%I_{IR}/I_M$. The values of I_{start} and $\%I_{IR}/I_M$ for curves (a)–(e) are shown in Table 2.28.

The data for curve (a) in Figure 2.34 show that the 15% I_{IR}/I_{IM} rule of thumb is applicable up to about 26% of the maximum TL intensity. Even at this high ratio of 26%, the activation energy obtained using the IR method is $E = 0.9964$ eV which corresponds to a very small error of 0.36% relative to the reference value of $E = 1$ eV used for the evaluation of the synthetic peak. The corresponding IR plot is shown in Figure 2.35.

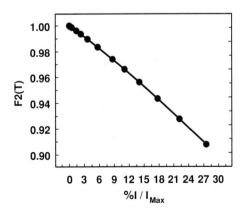

FIGURE 2.33. The values of the condition factor $F2(T)$ at various $\%I_{IR}/I_M$.

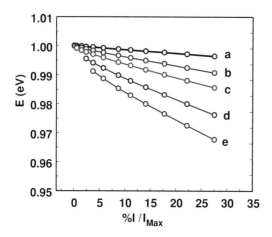

FIGURE 2.34. Activation energy values E resulting from the IR method by using the conditions given in the Table 2.28.

The results of Figure 2.34 show clearly that the commonly used 15% rule of thumb does not always apply. The correct application of the IR method depends strongly on the extent of the selected IR region. It also depends critically on the clear and accurate definition of the beginning of the IR region. In an experimental situation, this beginning must be defined according to the value of the background of the data, which also affects the results as was shown in the previous exercise.

Exercise 2.9: Error Analysis for Peak Shape Methods

Using a synthetic general-order glow-peak with $E = 1$ eV, $s = 10^{12}$ s^{-1}, and $b = 1.5$

(a) Find how the estimated error in the temperatures T_1, T_M, and T_2 is propagated in the symmetry factor μ, as a function of the error in temperature.

(b) Investigate what the error in temperature ΔT must be, so that one can discriminate between kinetic orders having a difference $\Delta b = 0.1$.

(c) Find how the estimated error in the temperatures T_1, T_M, T_2, and μ is propagated in the values of the activation energies E evaluated using general peak shape methods.

TABLE 2.28. I_{start} and T_{start} for the 15% rule of thumb study

Curve	T_{start} (K)	I_{start}/I_M
a	273	1.3×10^{-6}
b	314	3.13×10^{-4}
c	330	2.0×10^{-3}
d	346	1.0×10^{-2}
e	355	2.3×10^{-2}

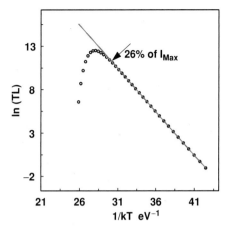

FIGURE 2.35. The open circles are the points on the glow peak and the solid line represents the IR line with slop $E = 1$ eV.

Solution

(a) The symmetry factor μ, is defined by the relation

$$\mu = \frac{T_2 - T_M}{T_2 - T_1}.\tag{2.21}$$

The errors on T_1, T_2, and T_M are propagated in the value of μ, according to the expression [5]:

$$\sigma_\mu = \pm\sqrt{\left(\frac{\partial\mu}{\partial T_1}\cdot\Delta T_1\right)^2 + \left(\frac{\partial\mu}{\partial T_2}\cdot\Delta T_2\right)^2 + \left(\frac{\partial\mu}{\partial T_M}\cdot\Delta T_M\right)^2}.\tag{2.22}$$

By evaluating the partial derivatives using equation (2.21) and substituting into equation (2.22), we find the following expression for the standard deviation σ_μ of μ:

$$\sigma_\mu = \pm\sqrt{\left(\frac{T_2 - T_M}{(T_2 - T_1)^2}\cdot\Delta T_1\right)^2 + \left(\frac{T_M - T_1}{(T_2 - T_1)^2}\cdot\Delta T_2\right)^2 + \left(-\frac{1}{T_2 - T_1}\cdot\Delta T_M\right)^2}\tag{2.23}$$

Using a spreadsheet program, one can easily generate a synthetic glow peak and evaluate the temperatures T_1, T_2, and T_M by using, for example, the series approximation from equation (1.52) in the general-order equation (1.7).

In experimental situations, the errors in temperature depend on the temperature interval used for sampling the TL signal. For example, if one measures the TL intensity per every 2 K, then the possible error in temperature T_M, T_1, T_2, etc., can be taken to be half of this interval (1 K).

In this exercise, the experimental situation is simulated by generating the synthetic glow peak using different temperature intervals to evaluate the TL intensity. This is shown in column 1 of Table 2.29. In the first row of the table, it is shown

TABLE 2.29. Data table 1 for error propagation in peak shape methods

Step	T_M (K)	T_1 (K)	T_2 (K)	ΔT (K)	μ	σ_μ	$3\sigma_\mu/\Delta\mu$
0.001	384.301	365.29	401.718	0.0005	0.478121	0.000017	0.005
0.005	384.305	365.29	401.715	0.0025	0.477968	0.000084	0.026
0.01	384.31	365.29	401.71	0.005	0.47759	0.00017	0.053
0.05	384.35	365.25	401.7	0.025	0.475995	0.00084	0.26
0.1	384.4	365.3	401.7	0.05	0.475275	0.00168	0.53
0.2	384.6	365.2	401.6	0.1	0.46033	0.00336	1.05
0.3	384.6	365.4	401.7	0.15	0.471074	0.005	1.57
0.4	384.6	365.4	401.8	0.2	0.472527	0.0067	2.1
0.5	385	365.5	401.5	0.25	0.465753	0.0085	2.67
0.6	384.6	365.4	402	0.3	0.475410	0.01	3.14
0.7	385	365.4	401.8	0.35	0.461538	0.012	3.52
0.8	385	365	401.8	0.4	0.456522	0.013	4.08
0.9	385.5	365.7	401.7	0.45	0.450000	0.015	4.71
1	385	365	402	0.5	0.459459	0.0165	5.18

that the first synthetic glow peak was generated by evaluating the TL every 0.001 K. The error in temperature is given by column 5 and was taken equal to one half of the step used in measuring TL ($\Delta T = 0.0005$ K).

For the sake of simplicity in the evaluations, it is assumed that $\Delta T_1 = \Delta T_2 = \Delta T_M = \Delta T$. The values of T_M, T_1, T_2, symmetry factor μ, and its standard deviation σ_μ according to equation (2.23) are listed in columns 2, 3, 4, 6, and 7 of Table 2.29, respectively, for different values of ΔT.

(b) From the first row of Table 2.29, the symmetry factor for $b = 1.5$ is equal to $\mu = 0.478121$. The symmetry factors for the neighboring values of $b = 1.4$ and $b = 1.6$ are found in a similar fashion to be equal to $\mu = 0.468176$ and $\mu = 0.487376$, respectively. The difference between these two symmetry factors for $b = 1.5$ and $b = 1.6$ is $\Delta\mu = 0.00925$.

In order for one to be able to discriminate between these two values, the following condition must be fulfilled.

$$3\sigma_\mu > \Delta\mu. \tag{2.24}$$

The data of Table 2.29 show that this condition holds only when $\Delta T > 0.1$ K. For all values of μ that satisfy the condition $3\sigma_\mu < \Delta\mu$, the respective values of μ_g belong to a kinetic order $b = 1.5$. Therefore, the error in temperature ΔT must be less than 0.2 K in order for the evaluated symmetry factor μ to correspond to the kinetic order $b = 1.5$, and not to 1.4 or 1.6.

(c) The general peak shape method equation for evaluating the activation energy E is given by equation (1.48):

$$E = c_\alpha \cdot \frac{kT_M^2}{a} - b_\alpha(2kT_M), \tag{1.48}$$

where α stands for τ, δ, and ω and the coefficients are given by equation (1.49):

$$c_\tau = 1.510 + 3.0(\mu - 0.42), \quad b_\tau = 1.58 + 4.2(\mu - 0.42)$$
$$c_\delta = 0.976 + 7.3(\mu - 0.42), \quad b_\delta = 0$$
$$c_\omega = 2.52 + 10.2(\mu - 0.42), \quad b_\omega = 1. \tag{1.49}$$

The errors on T_1, T_2, and T_M are propagated in the value of E, according to the expression:

$$\sigma_E = \pm \sqrt{\left(\frac{\partial E}{\partial T_M} \cdot \Delta T_M\right)^2 + \left(\frac{\partial E}{\partial a} \cdot \Delta a\right)^2 + \left(\frac{\partial E}{\partial \mu} \cdot \Delta \mu\right)^2}. \tag{2.25}$$

Taking into account equation (1.49), equation (2.25) becomes for the case of E_τ

$$\sigma_{E\tau} = \pm \sqrt{\left(\left(\frac{2c_\tau k T_M}{\tau} - 2kb_\tau\right) \cdot \Delta T_M\right)^2 + \left(\frac{c_\tau k T_M^2}{\tau^2} \cdot \Delta \tau\right)^2 + \left(\frac{3k T_M^2}{\tau} - 8.4k T_M\right) \cdot \Delta \mu)^2.4} \tag{2.26}$$

The corresponding equations for E_δ and E_ω are

$$\sigma_{E\delta} = \sqrt{\left(\frac{2c_\delta k T_M}{\delta} \cdot \Delta T_M\right)^2 + \left(\frac{c_\delta k T_M^2}{\delta^2} \cdot \Delta \delta\right)^2 + \left(\frac{7.3k T_M^2}{\delta} \cdot \Delta \mu\right)^2}, \tag{2.27}$$

$$\sigma_{E\omega} = \sqrt{\left(\frac{2c_\omega k T_M}{\omega} \cdot \Delta T_M\right)^2 + \left(\frac{c_\omega k T_M^2}{\omega^2} \cdot \Delta \omega\right)^2 + \left(\frac{10.2k T_M^2}{\omega} \cdot \Delta \mu\right)^2}. \tag{2.28}$$

For the sake of simplicity, we assume that $\Delta T_1 = \Delta T_2 = \Delta T_M = \Delta T$. The quantities τ, δ, and ω are given by the relations:

$$\omega = T_2 - T_1,$$
$$\delta = T_2 - T_M,$$
$$\tau = T_M - T_1. \tag{2.29}$$

The error propagation for ω is

$$\sigma_\omega = \pm\sqrt{\Delta T_2^2 + \Delta T_1^2}$$
$$\sigma_\omega = \pm \Delta T \sqrt{2} \tag{2.30}$$

with similar expressions for τ and δ. Therefore, in equations (2.26–2.28) one must use

$$\Delta \tau = \Delta \delta = \Delta \omega = 1.41 \, \Delta T.$$

The results for the error ΔE are shown in Table 2.30, which continues Table 2.29. The data in Table 2.30 show that the error in the activation energy depends on the peak shape quantity used in the evaluation (τ, δ, or ω). However, as in the previous question of this exercise, the most accurate values of the activation energy (with the lowest errors of 1% or less) are obtained when the temperature is measured with an accuracy better than 0.1 K.

TABLE 2.30. Data table 2 for error propagation in peak shape methods

ΔT	E_τ	$100 \times \sigma_{E\tau}$	E_δ	$100 \times \sigma_{E\delta}$	E_ω	$100 \times \sigma_{E\omega}$
0.0005	1.00672	0.0051	1.02316	0.0099	1.02125	0.0064
0.0025	1.0062	0.025	1.00624	0.0494	1.0208	0.0318
0.005	1.0056	0.051	1.00224	0.0989	1.02025	0.0635
0.025	1.9981	0.252	1.01598	0.495	1.01328	0.317
0.05	0.9971	0.504	1.01531	0.995	1.01247	0.636
0.1	0.9669	0.97	0.9982	2.02	0.9841	1.27
0.15	0.9852	1.49	1.0054	3.02	1.0015	1.92
0.2	0.9877	1.99	1.00742	4.00	1.00375	2.55
0.25	0.9489	2.4	0.9720	5.24	0.9664	3.25
0.3	1.0113	2.99	1.01125	5.91	1.0081	3.78
0.35	0.9488	3.33	0.9725	7.14	0.9666	4.45
0.4	0.9193	3.65	0.9447	8.05	0.9375	4.97
0.45	0.9214	4.19	0.9446	9.59	0.9388	5.84
0.5	0.924	4.56	0.9497	9.91	0.9425	6.15

References

[1] R. Chen and S.W.S. McKeever, 1997. *Theory of Thermoluminescence and Related Phenomena*. Singapore: World Scientific.
[2] G. Kitis, J.M. Gomez-Ros, and J.W.N. Tuyn, *J. Phys.* **D 31** (1998) 2636.
[3] C.M. Sunta, W.E.F. Ayta, T.M. Piters, and S. Watanabe, *Radiat. Meas.* **30** (1999) 197.
[4] R. Chen, D.J. Huntley, and G.W. Berger. *Phys. stat. sol. (a)* **79** (1983) 251.
[5] P.R. Bevington, 1969. *Data Reduction and Error Analysis for Physical Sciences*. New York: McGraw-Hill.

3
Simple TL Models

Introduction

In this chapter several examples are given of implementing simple thermoluminescence (TL) kinetic models. In the first two exercises of this chapter, the models are implemented in EXCEL using a simple numerical integration algorithm known as Euler's method. This method is known to be less accurate than the commonly employed higher order Runge–Kutta algorithms, but is included here for educational and demonstration purposes.

In the rest of this chapter several examples are given of using the programming language *Mathematica* to implement simple TL models. Some information on the algorithms and numerical integration techniques used in *Mathematica*, and a brief introduction to some of its features is given in the Appendix. The authors have found the numerical integration code in *Mathematica* to be very efficient and accurate, with most of the examples given in this chapter requiring typical running times of 1–2 min on a desktop computer. The complexity of the *Mathematica* programs is gradually increased in this chapter, and the reader is introduced incrementally to the capabilities of this powerful programming environment.

Exercise 3.3 demonstrates how *Mathematica* can be used to numerically integrate the differential equations for first-, second-, and general-order kinetics and how it can be used to present graphically the results of the integration. In Exercise 3.4 the differential equations for the well-known one-trap-one-recombination center model (OTOR) of TL are solved numerically, and the exact numerical solutions are compared with the commonly used quasi-equilibrium (QE) approximation. In Exercise 3.5 the kinetic equations for the more realistic and hence more complex interactive-multi-trap-system (IMTS) model for TL are developed, and detailed numerical results are presented.

In Exercise 3.6 the accuracy of an analytical expression for first-order kinetics is examined by calculating the figure of merit (FOM). This useful analytical expression for first-order kinetics is based on two experimentally measured parameters: the temperature T_M of maximum TL intensity, and the maximum TL intensity I_M. Exercises 3.7 and 3.8 are an extension of Exercise 3.6 for second- and general-order kinetics.

Exercise 3.9 is a comparative study of the accuracy of several analytical expressions available in the literature, which are used to represent first-order TL glow curves. Six different analytical expressions are compared and contrasted based on the goodness of fit to a simulated first order TL glow curve. Finally Exercise 3.10 is an extension of Exercise 3.9, consisting of a comparative study of five different analytical expressions for general order kinetics.

Finally Exercise 3.11 is a systematic study of the behavior of TL glow peaks exhibiting mixed order kinetics. A general comparison is made of mixed-order and general-order kinetics.

Whenever possible, special effort has been made to use the same terms and symbols as in the original research papers in the literature, in order to facilitate the work of researchers interested in pursuing more detailed studies of the numerical models in this chapter.

Exercise 3.1: Numerical Integration of First-Order Equation

Use a calculator or a spreadsheet program like EXCEL to integrate numerically the standard equation for first-order kinetics

$$I(t) = -\frac{dn}{dt} = nse^{-E/kT},$$ (3.1)

where E = thermal activation energy of the trap (eV)
s = frequency factor (s^{-1})
T = temperature of the sample (K)
k = Boltzmann constant (eV K^{-1})
N = total concentration of the traps in the crystal (cm^{-3})
n = filled concentration of the traps in the crystal (cm^{-3})
n_0 = initial concentration of filled traps at time $t = 0$ (cm^{-3}).

Use a simple numerical integration method such as Euler's method for numerical integration [1]. The program should allow the user to change the values of the parameters E, s, β and n_0 in order to examine the effect of these parameters on the TL glow curve and should graph the numerically obtained graphs TL(T), $n(T)$.

Use the following set of numerical values of the parameters: $n_0 = 10^{10}$ cm^{-3}, $E = 1$ eV, $s = 10^{12}$ s^{-1}, $k = 8.617 \times 10^{-5}$ eV K^{-1}, $\beta = 1$ K s^{-1} and assume a linear heating rate β.

Solution

The differential equation can be written in terms of the temperature T instead of the time t by writing

$$\frac{dn}{dt} = \frac{dn}{dT}\frac{dT}{dt} = \frac{dn}{dT}\beta = -nse^{-E/kT},$$ (3.2)

where β = linear heating rate in K s^{-1}.

By writing $\dfrac{dn}{dT} = \dfrac{\Delta n}{\Delta T}$ and rearranging the given differential equation we obtain

$$\Delta n = n(T + \Delta T) - n(T) = -nse^{-E/kT}\frac{\Delta T}{\beta}. \tag{3.3}$$

The left-hand side of this equation represents the difference between the value of the electron concentration at temperatures T and $T + \Delta T$, so that by rearranging:

$$n(T + \Delta T) = n(T) - n(T)se^{-E/kT}\frac{\Delta T}{\beta}. \tag{3.4}$$

This equation forms the basis of the integration procedure we will use to numerically integrate the first-order equation and is known as Euler's method. The calculations can be carried out easily using a calculator, or by using a spreadsheet.

By using the values of $n_0 = 10^{10}$ cm^{-3}, $E = 1$ eV, $s = 10^{12}$ s^{-1}, $k = 8.617 \times 10^{-5}$ eV K^{-1}, $\Delta T = 1$ K, $\beta = 1$ K s^{-1} we obtain the following values of $n(T)$ using equation (3.4):

$$T_0 = 20°C = 293 \text{ K}$$
$$n_0 = 10^{10}$$
$$T_1 = 21°C = 294 \text{ K}$$

$$n_1 = n_0 - n_0 se^{-E/kT_1}\frac{\Delta T}{\beta} = 10^{10} - 10^{10} \times 10^{12}\, e^{-\frac{1.0}{8.617 \times 10^{-5} \times 294}}\frac{1}{1}$$

$$= 10^{10} - 71984$$

$$TL = -\beta\frac{\Delta n}{\Delta T} = \beta\frac{n_0 - n_1}{\Delta T} = 1\frac{10^{10} - (10^{10} - 71984)}{1} = 71984$$

$$T_2 = 22°C = 295 \text{ K}$$

$$n_2 = n_1 - n_1 se^{-E/kT_2}\frac{\Delta T}{\beta}$$

$$= 10^{10} - (10^{10} - 71984) \times 10^{12} e^{-\frac{1.0}{8.617 \times 10^{-5} \times 295}}\frac{1}{1} = 10^{10} - 82289$$

$$TL = -\beta\frac{\Delta n}{\Delta T} = \beta\frac{n_0 - n_1}{\Delta T} = 1\frac{10^{10} - (10^{10} - 82289)}{1} = 82289.$$

By continuing in a similar manner, we can obtain Table 3.1 with the first five values of $n(T)$.

The above series of calculations is best implemented on a spreadsheet type of program like EXCEL. The EXCEL program will contain three columns: the first column contains the temperature values T at intervals of $\Delta T = 1°C$, while the second column will contain the values of $n(T + \Delta T)$ calculated from the above equation. The third column will contain the difference between successive values of the concentrations $n(T)$, and will be proportional to the observed TL intensity.

TABLE 3.1. Calculation of $n(T)$ for first-order
kinetics equation

$T(^\circ C)$	$n(T)$	$TL(T) = -\beta dn/dT$
20	10,000,000,000	
21	9,999,928,016	71,984
22	9,999,845,727	82,289
23	9,999,751,742	93,985
24	9,999,644,496	107,246

The result of implementing the above calculations in an EXCEL spreadsheet is shown in Table 3.2, together with the graphs of $n(T)$ and $TL(T)$ in Figure 3.1. For the sake of brevity, only the first few lines of the spreadsheet calculations are shown here. The following formulas are entered in cells C6, C7 and D7 in order to implement Euler's equation.

Cell C6 = $B$3,

CellC7 = C6-C6*$B$1*EXP(−$B$2/(0.00008617*(273 + B7)))*1,

CellD7 = C6-C7, etc.

An important advantage of using a spreadsheet is that by changing the value of one of the parameters in the spreadsheet (such as s, E, n_0), the user can immediately see the resulting changes in the graphs of $n(T)$ and $TL(T)$.

Exercise 3.2: The One-Trap-One-Recombination Center Model

Use EXCEL to integrate numerically the differential equation describing the TL process for the OTOR model. This model consists of an isolated electron trap and a recombination center (RC), as shown in Figure 3.2.

TABLE 3.2. Calculations of $n(T)$ and $TL(T)$ for first-order
kinetics equation

	A	B	C	D
1	$s =$	1.00×10^{12}	s^{-1}	
2	$E =$	1	eV	
3	$n_0 =$	1.00×10^{10}	cm^{-3}	
4				
5		$T(^\circ C)$	$n(T)$	$TL(T) = -\beta dn/dT$
6		20	10,000,000,000	
7		21	9,999,928,016	71,984
8		22	9,999,845,727	82,289
9		23	9,999,751,742	93,985
10		24	9,999,644,496	107,246

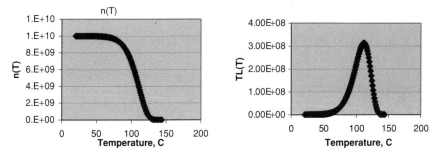

FIGURE 3.1. Results from integrating the equation for first-order kinetics.

By applying the QE conditions, we arrive at the following general analytical expression for the TL intensity in the OTOR model [2]:

$$I = -\frac{dn}{dt} = \frac{sn^2 \exp\left(-\dfrac{E}{kT}\right)}{nA_h + (N - n)A_n} \cdot A_h, \qquad (3.5)$$

where E = thermal activation energy of the trap (eV)

s = frequency factor (s^{-1})

T = temperature of the sample (K)

k = Boltzmann constant (eV K^{-1})

N = total concentration of the traps in the crystal (cm^{-3})

n = concentration of filled traps in the crystal (cm^{-3})

n_0 = initial concentration of filled traps at time $t = 0$ (cm^{-3})

A_n = probability coefficient of electron retrapping in the traps (cm^3 s^{-1})

A_h = probability coefficient of electron recombining with holes in the RC (cm^3 s^{-1}).

Use Euler's method in a spreadsheet. The program should allow the user to change the values of E, s, β, n_0, A_n, and A_h in order to see the effect of these parameters on the TL glow curve, and should graph the numerically obtained graphs TL(T), $n(T)$.

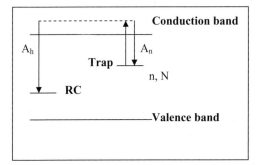

FIGURE 3.2. The OTOR model.

Use the following values for the parameters:

$$E = 1 \text{ eV}, s = 10^{12} \text{ s}^{-1}, N = 10^{10} \text{ cm}^{-3}, A_n = 10^{-7} \text{ cm}^3 s^{-1},$$
$$\beta = 1 \text{ Ks}^{-1}, A_n/A_h = 10^{-2}, n_0 = N$$

The conditions under which the OTOR model produces first-order or second-order TL glow curves can be examined by varying the above parameters.

Show that with the given parameters, the OTOR model produces a first-order TL glow peak. Since in our case $A_n/A_h = 10^{-2}$, the probability coefficient of retrapping in the trap is much smaller than the recombination probability coefficient of the electrons in the RC, and we would expect the OTOR model to produce a first-order TL peak.

Solution

The given equation can be written as

$$\frac{dn}{dt} = -\frac{sn^2 \exp\left(-\dfrac{E}{kT}\right)}{nA_h + (N - n)A_n} \cdot A_h. \tag{3.6}$$

As in the previous example, the derivative dn/dt can be written in terms of the temperature T instead of the time t by writing

$$\frac{dn}{dt} = \frac{dn}{dT}\frac{dT}{dt} = \frac{dn}{dT}\beta, \tag{3.7}$$

where β = linear heating rate in $K\,s^{-1}$.

By approximating $\dfrac{dn}{dt} = \dfrac{\Delta n}{\Delta T}$ and rearranging we obtain

$$\Delta n = n(T + \Delta T) - n(T) = -\frac{sn^2 \exp\left(-\dfrac{E}{kT}\right)}{nA_h + (N - n)A_n} \cdot A_h \cdot \frac{\Delta T}{\beta}. \tag{3.8}$$

Finally by rearranging:

$$n(T + \Delta T) = n(T) - \frac{sn^2 \exp\left(-\dfrac{E}{kT}\right)}{nA_h + (N - n)A_n} \cdot A_h \cdot \frac{\Delta T}{\beta}. \tag{3.9}$$

As in the previous example, we will use this equation which expresses Euler's numerical integration method to numerically integrate the differential equation.

We set up an EXCEL spreadsheet as shown in Table 3.3. The first 10 rows of the spreadsheet contain the input parameters of the OTOR model. Only the first five calculated values of $I(T)$ and $n(T)$ are shown. Column A contains the values of time $t = 0,1,2.$ s and column B contains the corresponding temperature $T = T_0 + \beta t$ where $T_0 = 21°C$ = temperature at time $t = 0$ and β = heating rate = $1 K\,s^{-1}$.

TABLE 3.3. Calculations of $n(T)$ and $TL(T)$ for the OTOR model

	A	B	C	D
1	OTOR model of Thermoluminescence			
2	Input parameters			
3	$E =$	1	eV	
4	$s =$	1.00×10^{12}	s^{-1}	
5	$N =$	1.00×10^{10}	cm^{-3}	
6	$A_n =$	1.00×10^{-7}	$cm^3 s^{-1}$	
7	$A_h =$	1.00×10^{-5}	$cm^3 s^{-1}$	
8	$\beta =$	1	$°Cs^{-1}$	
9	$n_0 =$	1.00×10^{10}	cm^{-3}	
10	$\Delta T =$	1	$°C$	
11				
12	time t (s)	Temperature T ($°C$)	$n(T)$	TL
13	0	21	1.00×10^{10}	
14	1	22	9999917710.31	82289.69
15	2	23	9999823724.94	93985.37
16	3	24	9999716477.76	107247.18
17	4	25	9999594206.00	122271.76
18	5	26	9999454927.25	139278.76

Column C contains the calculation for the instantaneous concentration of filled traps $n(T)$. The calculation is carried by using the formulas:

Cell C13 $=$ $B$9 (this is the initial concentration n_0),

Cell C14 $=$ C13-(C13*C13*$B$7*$B$4*EXP(-$B$3/(0.00008617*

(273 + B14)))/$B$8)/(($B$5-C13)*$B$6 + C13*$B$7),

Cell C15 $=$ C14 $-$ (C14*C14*$B$7*$B$4*EXP(-$B$3/(0.00008617*

(273 + B15)))/$B$8)/(($B$5-C14)*$B$6 + C14*$B$7), etc.

Column D contains the calculated values of the TL intensity:

$$D14 = C13\text{-}C14$$
$$D15 = C14\text{-}C15 \quad \text{etc.}$$

The results of the calculation using Euler's method are shown in Table 3.3 and graphs of TL(T), $n(T)$ are shown in Figure 3.3.

Since for the given parameters $A_n/A_h = 10^{-2}$, the retrapping probability of the thermally ejected electrons is much smaller than the probability of their recombining at the RC, we might expect the calculated TL glow peak to follow first-order kinetics. The calculated TL glow curve in Figure 3.3 has indeed the characteristic asymmetric shape of first-order kinetics peaks, with the following temperature

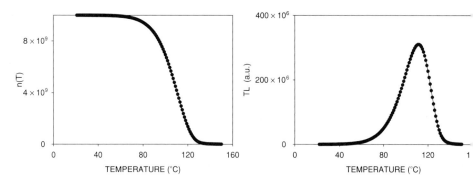

FIGURE 3.3. The results from the OTOR model.

parameters:

$$T_M = 384 \text{ K}$$
$$\tau = 17 \text{ K}$$
$$\delta = 13 \text{ K}$$
$$\omega = 30 \text{ K}$$

and the geometrical shape factor $\mu = \delta/\omega = 13/30 = 0.43$.

As expected, the geometrical shape factor is very close to the expected value of $\mu = 0.42$ for a first-order kinetics peak.

Several more examples of results obtained with the OTOR model are given in Exercise 3.4 of this chapter by using *Mathematica*.

Exercise 3.3: Calculation of Glow Peaks Using Mathematica

(a) Write a computer program in *Mathematica* to numerically integrate the equations for first-, second- and general-order kinetics:

$$I(t) = -\frac{dn}{dt} = nse^{-E/kT}$$

$$I(t) = -\frac{dn}{dt} = \frac{n^2}{N}se^{-E/kT}$$

$$I(t) = -\frac{dn}{dt} = n^b\frac{s}{N}e^{-E/kT} \qquad (3.10)$$

Use the following numerical values: $E = 1.0$ eV, $s = 10^{12}$ s^{-1}, $\beta = 1$ K s^{-1}, $n_0 = N = 10^{10}$ cm^{-3}.

(b) Modify the program so that it graphs the solutions for different initial electron trap occupancies n_0. Show that in the case of first-order kinetics, changes in n_0 do not affect the position of the maximum intensity (T_{max}) of the first-order TL glow curve, but these changes affect the maximum height (I_{max}) of the glow curve.

(c) Show that in the cases of second- and general-order kinetics, changes in the initial electron trap concentration n_0 affect the position of both the maximum intensity (T_{max}) and the maximum height (I_{max}) of the glow curve.

Solution

The following simple program in *Mathematica* solves the differential equation for first-order kinetics with the initial value $n_0 = N$. The parameter **beta1** represents the linear heating rate β. The command **Remove** is always introduced at the beginning of all *Mathematica* programs in this book, in order to clear the memory from all variables and programs. The command **NDSolve** is used to perform the numerical integration of the differential equation, and stores the result of the numerical integration as the parameter **sol** (which stands for the **sol**ution of the differential equation). The parameter **x** represents the temperature T (in °C) and the function $n1[x]$ represents the electron concentration $n(T)$. The integration is carried from a temperature of $T = 0$°C to $T = 200$°C, and a maximum number of integration steps is set at 50,000 by using the variable **MaxSteps**. The command **Plot** is used to graph the result $n(T)$ and TL(T) of the numerical integration procedure.

(a) First-Order Kinetics

```
Remove["Global`*"];
E1 = 1.0; s1 = 10^12; k1 = 8.617*10^-5; beta1 = 1;
  N1 = 10^10; no = 1*N1; Npoints = 200;
sol = NDSolve[{n1'[x] == -n1[x]*(s1/beta1)*E^(-E1/
  (k1*(273+x))), n1[0] == no}, {n1}, {x, 0, Npoints},
  MaxSteps → 50000];
Plot[Evaluate[n1[x]/.sol], {x,0,Npoints}, PlotRange →
  All, AxesLabel → {"T", "n(T)"}];
Plot[Evaluate[-n1'[x]/.sol],{x,0,Npoints}, PlotRange → All,
  AxesLabel → {"T", "TL(T)"}];
```

The results of running the program for first order kinetics are given in Figure 3.4. By changing the third line containing the parameter **sol** in the above simple program, we obtain the solution of the second-order and general-order equations as follows.

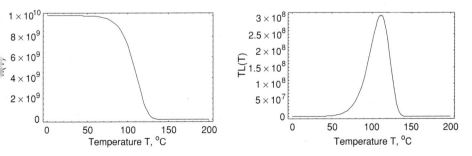

FIGURE 3.4. The result of numerical integration for first-order kinetics using *Mathematica*.

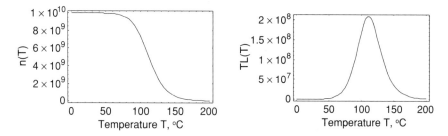

FIGURE 3.5. The result of numerical integration for second-order kinetics using *Mathematica*.

The results of integrating the second-order kinetic equation are shown in Figure 3.5.

Second-Order Equation

```
Remove["Global`*"];
El = 1.0; s1 = 10^12; k1 = 8.617*10^-5; beta1 = 1;
  N1 = 10^10; no = 1*N1; Npoints = 200;
sol = NDSolve[{n1'[x] == -n1[x]^2/N1*(s1/beta1)*E^(-El/
  (k1*(273+x))), n1[0] == no}, {n1}, {x, 0, Npoints},
  MaxSteps → 50000];
Plot[Evaluate[n1[x]/.sol],{x,0,Npoints}, PlotRange → All,
  AxesLabel → {"T", "n(T)"}];
Plot[Evaluate[-n1'[x]/.sol],{x,0,Npoints}, PlotRange → All,
  AxesLabel → {"T", "TL(T)"}];
```

General-Order Kinetics

By simply changing the third line and by increasing the number of points **Npoints** in the previous program, the following integrates the differential equation for general-order kinetics with a general-order value of $b = 1.5$.

The results of integrating the general-order kinetics equation are shown in Figure 3.6.

```
Remove["Global`*"];
El = 1.0; s1 = 10^12; k1 = 8.617*10^-5; beta1 = 1;
  N1 = 10^10; no = 1*N1; Npoints = 450; b = 1.5;
sol = NDSolve[{n1'[x] == -n1[x]^b/N1*(s1/beta1)*E^(-El/
  (k1*(273+x))), n1[0] == no}, {n1}, {x, 0, Npoints},
  MaxSteps → 50000];
Plot[Evaluate[n1[x]/.sol], {x,0,Npoints}, PlotRange → All,
  AxesLabel → {"T", "n(T)"}];
Plot[Evaluate[-n1'[x]/.sol],{x,0,Npoints}, PlotRange → All,
  AxesLabel → {"T", "TL(T)"}];
```

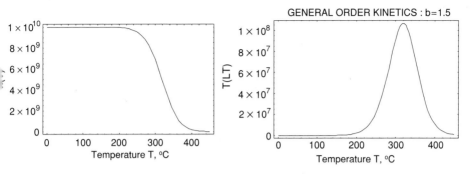

FIGURE 3.6. The result of numerical integration for general-order kinetics using *Mathematica*.

(b) Graphing Different Values of the Parameter n_0—First-Order Kinetics

By making simple changes in the program we can graph the solutions of the first-order kinetics for three different values of the parameter $n_0 = N, 0.5N,$ and $0.1N$. The solutions of the first-order equation for these three values of n_0 are stored in the variables **sol1, sol2, sol3** and are graphed using the **Plot** command. The graphs are stored in the graphic objects **gr1, gr2, and gr3.** Finally the three graphs are shown together by using the *Mathematica* command **Show**.

```
Remove["Global`*"];
E1 = 1.0; s1 = 10^12; k1 = 8.617*10^-5; beta1 = 1;
N1 = 10^10; no = 1*N1; Npoints = 200;
sol1 = NDSolve[{n1'[x] == -n1[x]*(s1/beta1)*E^(-E1/
  (k1*(273+x))), n1[0] == no}, {n1}, {x, 0, Npoints}];
no = 0.5*N1;
sol2 = NDSolve[{n1'[x] == -n1[x]*(s1/beta1)*E^(-E1/
  (k1*(273+x))), n1[0] == no}, {n1}, {x,0, Npoints}];
no = 0.1*N1;
sol3 = NDSolve[{n1'[x] == -n1[x]*(s1/beta1)*E^(-E1/
  (k1*(273+x))), n1[0] == no}, {n1}, {x, 0, Npoints}];
gr1 = Plot[Evaluate[-n1 '[x] /.sol1], {x, 0, Npoints},
  AxesLabel -> {"T", "TL(T)"}, PlotRange -> All];
gr2 = Plot[Evaluate[-n1 '[x] /.sol2], {x, 0, Npoints},
  AxesLabel -> {"T", "TL(T)"}, PlotRange -> All];
gr3 = Plot[Evaluate[-n1 '[x] /.sol3], {x, 0, Npoints},
  AxesLabel -> {"T", "TL(T)"}, PlotRange -> All];
Show[{gr1, gr2, gr3}, PlotLabel -> "First order kinetics:
  no=1*N, 0.5*N, 0.1*N", ImageSize -> 755];
```

As seen in Figure 3.7, the position of the maximum intensity (T_{max}) of the first-order TL glow curve stays the same for different values of the parameter n_0, but the maximum height (I_{max}) of the glow curve decreases, while the overall asymmetric

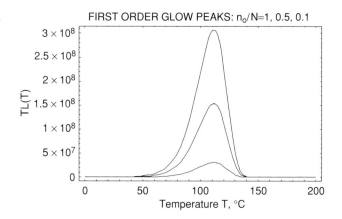

FIGURE 3.7. First-order kinetics calculation for different values of the parameter n_0.

shape of the TL glow curve remains the same. The symmetry factor $\mu = 0.42$ for all graphs is shown in Figure 3.7.

(c) Graphing Different Electron Trap Occupancies n_0—Second Order Kinetics

By making simple changes in the program we can also graph the solutions of the second-order kinetics for three different electron trap occupancies $n_0 = N$, 0.5N, and 0.1N. The results are shown in Figure 3.8.

```
Remove["Global`*"];
E1 = 1.0; s1 = 10^12; k1 = 8.617*10^-5; beta1 = 1;
N1 = 10^10; no = 1*N1; Npoints = 200;
sol1 = NDSolve[{n1'[x] == -n1[x]^2/N1*(s1/beta1)*
  E^(-E1/(k1*(273+x))), n1[0] == no}, {n1},
  {x, 0, Npoints}];
no = 0.5*N1;
sol2 = NDSolve[{n1'[x] == -n1[x]^2/N1*(s1/beta1)*
  E^(-E1/(k1*(273+x))), n1[0] == no}, {n1},
  {x, 0, Npoints}];
no = 0.1*N1;
sol3 = NDSolve[{n1'[x] == -n1[x]^2/N1*(s1/beta1)*
  E^(-E1/(k1*(273+x))), n1[0] == no}, {n1},
  {x, 0, Npoints}];
gr1 = Plot[Evaluate[-n1 '[x] /.sol1], {x, 0, Npoints},
  AxesLabel -> {"T", "TL(T)"}, PlotRange -> All];
gr2 = Plot[Evaluate[-n1 '[x] /.sol2], {x, 0, Npoints},
  AxesLabel -> {"T", "TL(T)"}, PlotRange -> All];
gr3 = Plot[Evaluate[-n1 '[x] /.sol3], {x, 0, Npoints},
  AxesLabel -> {"T", "TL(T)"}, PlotRange -> All];
Show[{ gr1, gr2, gr3}, PlotLabel -> "Second order kinetics:
  no=1*N, 0.5*N, 0.1*N", ImageSize -> 755];
```

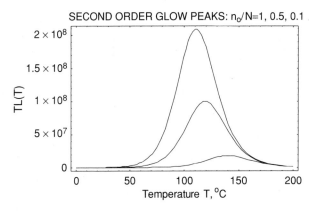

FIGURE 3.8. Second-order kinetics calculation for different initial occupancies.

As seen in Figure 3.8, both the position of the maximum intensity (T_{max}) of the second-order TL glow curve and the maximum height (I_{max}) of the glow curve change with the value of n_0. Nevertheless, the geometrical shape factor remains the same ($\mu = 0.52$) for all graphs in Figure 3.8.

Figure 3.9 shows the general-order graphs for three different electron trap occupancies $n_0/N = 1, 0.5,$ and 0.1 and for a kinetic order $b = 1.5$. The rest of the parameters are the same as in the case of first- and second-order kinetics.

As seen in Figure 3.9, both the position of the maximum intensity (T_{max}) and the maximum height (I_{max}) of the general-order TL glow curve change with the value of n_0. Nevertheless, the overall shape of the glow peaks remains the same for all graphs in Figure 3.9.

Exercise 3.4: The OTOR Model in Mathematica

(a) The OTOR model was introduced in Exercise 3.2 of this chapter. A detailed study of the OTOR model can be found in Ref. [4], and in its most general form

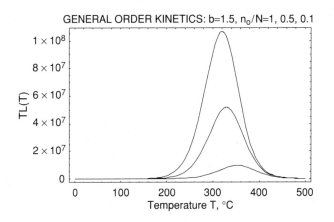

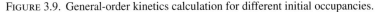

FIGURE 3.9. General-order kinetics calculation for different initial occupancies.

the OTOR is described by the following equations:

$$\frac{dn}{dt} = -sn \exp\left(-\frac{E}{kT}\right) + (N - n)A_h n_c \tag{3.11}$$

$$\frac{dn_c}{dt} = -\frac{dn}{dt} - A_h n_c(n + n_c) \tag{3.12}$$

$$\frac{dn_h}{dt} = \frac{dn}{dt} + \frac{dn_c}{dt} \tag{3.13}$$

$$I_{TL} = -\frac{dn_h}{dt} = n_c(n + n_c)A_h. \tag{3.14}$$

All symbols in these equations are identical to the ones employed in Exercise 3.2 of this chapter. In addition, $n_c(t)$ and $n_h(t)$ represent the instantaneous concentrations of the electrons and holes in the conduction band and RC, respectively.

The first equation expresses mathematically the fact that electrons in the trap can be either thermally excited in the conduction band (term $-sn\exp(-E/kT)$), or they can be retrapped in the trap with a probability coefficient A_n (term $(N - n)A_n n_c$). The second equation represents the change in the concentration of the electrons in the conduction band n_c. The concentration of these free electrons in the conduction band can be reduced by either trafficking into the trap (term $-dn/dt$), or by recombining in the RC with a probability coefficient A_h (term $-A_h n_c(n + n_c)$).

The third equation describes the conservation of total charge in the crystal, with the left-hand side being equal to the rate of change of the concentration of holes trapped in the RC, and the right-hand side representing the rate of change of the total concentration of electrons in the crystal.

Write a *Mathematica* program that solves this system of differential equations for the OTOR model. Use the same numerical values as in Exercise 3.2 in this chapter.

(b) Compare the exact solution of the OTOR differential equations from part (a) with the following approximate equation derived using the QE approximation [4]:

$$I_{QE}(T) = -\frac{dn}{dt} = \frac{sn^2 \exp\left(-\dfrac{E}{kT}\right)}{nA_h + (N - n)A_n} \cdot A_h. \tag{3.15}$$

Examine the accuracy of the quasistatic equilibrium approximation, by graphing the exact solution from part (a) together with the above expression (3.15) for $I_{QE}(T)$ on the same graph.

(c) Show that in the case of $A_n/A_h = 10^{-2}$ and $n_0 = N$, the OTOR model produces a first-order TL peak. This corresponds to the situation where the probability of retrapping of electrons in the trap is much smaller than the recombination probability of the electrons in the RC.

In the case $A_n/A_h = 100$ (probability coefficient of retrapping is much larger than the recombination probability coefficient) and $n_0 = N$, show that the OTOR model produces a second-order TL peak.

In the case of $A_n/A_h = 1$ and $n_0 = N$ (equal retrapping and recombination probabilities), show that the OTOR model produces a second-order TL peak.

Solution

(a) The following *Mathematica* program integrates the differential equations for the OTOR model. The solution is stored in the parameter **sol** as in previous examples, and the command **Plot** is used to graph the functions nc[x], n1[x], nh[x], -nh'[x] which correspond to the functions $n_c(T)$, $n(T)$, $n_h(T)$, and TL(T). The graph of the TL glow curve is stored in the graphical object **gr1**.

In the second part of the program the differential equation (3.15) is integrated and the result is stored in the parameter **solQE**. The command **Plot** is used to graph the function $-$nh'[x] which represents the intensity of the TL glow curve. The corresponding graph of the TL glow curve is stored in the object **gr2**, and the command **Show** is used to graph both the solutions of (3.15) and of the system (3.11)–(3.14) on the same graph.

```
Remove["Global`*"];
E1 = 1.0; s1 = 10^12; k1 = 8.617*10^-5; beta1 = 1;
N1 = 10^10; An = 10^-7; Ah = 100*An; no = 1*N1;
  Npoints = 200;
sol = NDSolve[{n1'[x] == -n1[x]*(s1/beta1)*E^(-E1/
  (k1*(273+x)))
  +An*(N1-n1[x])*nc[x]/beta1, nc'[x] == -n1'[x]-Ah*nc[x]*
  (n1[x]+nc[x])/beta1, nh'[x] == n1'[x]+nc'[x],
  n1[0] == no, nc[0] == 0, nh[0] == n1[0]+nc[0],
  {n1, nc, nh}, {x, 0, Npoints}];
Plot[Evaluate[n1[x]/.sol], {x, 0, Npoints},
  AxesLabel -> {"T","n(T)"}];
Plot[Evaluate[nc[x]/.sol], {x, 0, Npoints},
  AxesLabel -> {"T", "nc(T)"}];
Plot[Evaluate[nh[x]/.sol], {x, 0, Npoints},
  AxesLabel -> {"T", "nh(T)"}];
gr1 = Plot[Evaluate[-nh'[x]/.sol], {x, 0, Npoints},
  AxesLabel -> {"T", "TL(T)"}];
Print["OTOR:E=", N[E1], "s=", N[s1], "β=", N[beta1], "no=",
  N[no], "N=", N[N1], "Ah=", N[Ah],  "An=", N[An]];
solQE = NDSolve[{n2'[x] == -n2[x]^2*(s1/beta1)*E^(-E1/
  (k1*(273+x)))*Ah/(n2[x]*Ah+(N1-n2[x])*An),
  n2[0] == no}, {n2}, {x, 0, Npoints}];
gr2 = Plot[Evaluate[-n2'[x]/.solQE], {x, 0, Npoints},
  AxesLabel -> {"T", "TLQE(T)"}];
Show[{gr1, gr2}, AxesLabel -> {"T", "TL(T) and TLQE(T)"}];
Plot[Evaluate[(-nh'[x]/.sol)+(n2'[x]/.solQE)],
  {x, 0, Npoints}, AxesLabel -> {"T", "TL(T)-TLQE(T)"},
  PlotRange -> All];
```

The results from running the program are shown in Figure 3.10.

The results of Figure 3.10 show that as the sample is heated from 0 to 200°C, the concentrations of trapped electrons and holes $n(T)$ and $n_h(T)$ decrease

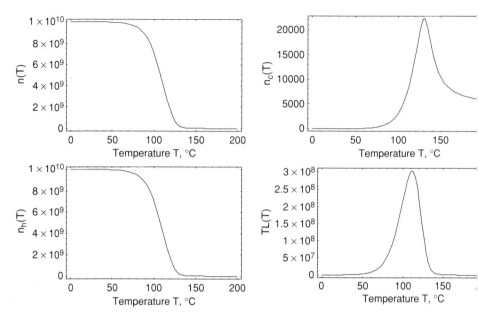

FIGURE 3.10. Results of the OTOR model in *Mathematica*.

simultaneously, with the maximum decrease taking place around 110°C. This temperature coincides with the observed maximum of the TL glow curve.

The graph $n_c(T)$ shows that the concentration of free electrons in the conduction band increases up to a temperature of approximately 130°C, which is higher than the corresponding maximum of the glow curve. This result means that one would expect the TL and thermally stimulated current curve (TSC) to occur at different temperatures. This is indeed found experimentally for some dosimetric materials.

(b) The last four lines in the program calculate the TL intensity $I_{QE}(T)$ as given by the quasistatic equilibrium approximation in equation (3.15). The result of graphing together the exact solution $I(T)$ from part (a), and the approximation $I_{QE}(T)$ from equation (3.15) are shown in Figure 3.11. As can be seen, the approximation result is indistinguishable from the exact solution, at least on the scale of the graph shown here.

Figure 3.11 shows the difference $I(T) - I_{QE}(T)$ as a function of the temperature T. It can be seen that the differences between the two intensities are indeed very small at all temperatures (with the residuals being less than 0.001% of the TL intensity).

(c) The graphs in Figure 3.12 show the results from the program using different A_n/A_h ratios. The rest of the parameters in the computer program are left the same.

In Figure 3.12(a) the ratio $A_n/A_h = 0.01$, i.e. the probability coefficient A_h of the free electrons in the conduction band recombining in the RC is 100× larger than the probability coefficient A_n of being retrapped in the electron trap. One would expect under these circumstances that first-order kinetics will dominate the kinetic process: The calculated TL glow curve has indeed the characteristic shape for first-order peaks.

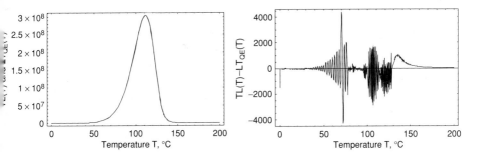

FIGURE 3.11. Comparison of the exact solution for the OTOR model with the quasi-equilibrium (QE) approximation.

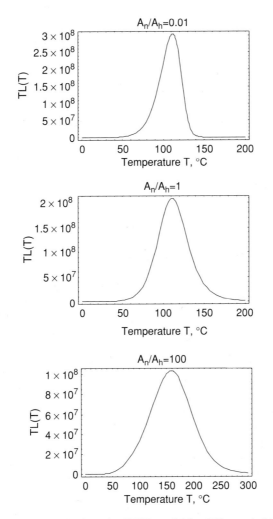

FIGURE 3.12. The results from the OTOR model for different A_n/A_h values.

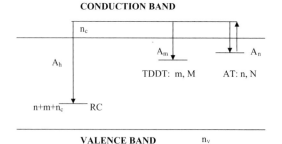

FIGURE 3.13. The IMTS model.

In Figure 3.12(b) the ratio $A_n/A_h = 1$, i.e. the probability coefficient A_h of recombining in the RC is equal to the retrapping probability coefficient A_n. One would expect under these circumstances that second-order kinetics will dominate the kinetic process: The calculated TL glow curve has indeed the characteristic shape for second-order peaks. By calculating the geometrical shape factor μ in Figure 3.12(b) one obtains $\mu = 0.52$ and hence this glow curve has the shape of second-order kinetics peak.

In Figure 3.12(c) the ratio $A_n/A_h = 100$, i.e. the probability coefficient A_h of recombining in the RC is $100\times$ smaller than the retrapping probability coefficient A_n. One would expect under these circumstances that heavy retrapping will affect the shape of the glow curve, by extending the observed TL peak over a much larger temperature range. This is seen to be indeed the case in Figure 3.12(c).

It must be noted that the situation $A_n/A_h = 100$ shown in Figure 3.12(c) represents a purely theoretical result and has not been observed in actual experimental results.

Exercise 3.5: The IMTS Model in Mathematica

(a) Write a computer program to integrate the kinetic rate equations relevant to the IMTS model shown in Figure 3.13 [2]. The model consists of two trapping states and a RC. The first trap is considered to be the one responsible for TL (denoted as the active trap (AT) below), and the second trap is denoted as the thermally disconnected deep trap (TDDT). The two traps are characterized by total concentrations N and M, and by instantaneous occupancies $n(t)$ and $m(t)$, respectively.

Because of the conservation of charge in the crystal, the concentration of holes in the RC at any moment must be equal to the total instantaneous concentration of electrons $n + m + n_c$.

The kinetic equations for this model are [2]

$$\frac{dn}{dt} = -ns \, \exp(-E/kT) + A_n(N - n)n_c \qquad (3.16)$$

$$\frac{dm}{dt} = A_m(M - m)n_c \qquad (3.17)$$

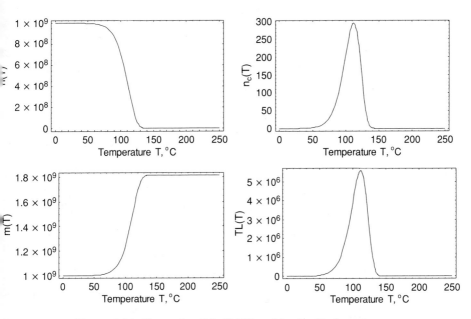

FIGURE 3.14. The results of the IMTS model using *Mathematica*.

$$\frac{dn_c}{dt} + \frac{dn}{dt} + \frac{dm}{dt} = -(n + m + n_c)n_c A_h \tag{3.18}$$

$$I = -\frac{d(n + m + n_c)}{dt} = (n + m + n_c)n_c A_h, \tag{3.19}$$

where E = thermal activation energy of the trap (eV)
    s = frequency factor (s^{-1})
    T = temperature of the sample (K)
    k = Boltzmann constant $(eV\ K^{-1})$
    N = total concentration of AT in the crystal (cm^{-3})
    n = instantaneous concentration of filled AT in the crystal (cm^{-3})
    n_0 = initial concentration of filled traps at time $t = 0$ (cm^{-3})
    A_n = probability coefficient of electron retrapping in the AT traps $(cm^3\ s^{-1})$
    A_h = probability coefficient of electron recombination with holes in the RCs $(cm^3 s^{-1})$
    A_m = probability coefficient of electron retrapping in the TDDT traps $(cm^3 s^{-1})$
    m = concentration of filled TDDT traps in the crystal (cm^{-3})
    M = total concentration of available TDDT traps in the crystal (cm^{-3})
    n_c = concentration of electrons in the conduction band (cm^{-3})
$n + m + n_c$ = total instantaneous concentration of electrons and holes in the crystal (cm^{-3}).

The interpretation of the first equation is identical to the one given in the previous example for the OTOR model. The second equation expresses the fact that electrons can be trapped into the TDDT with a probability coefficient A_m (term $A_m(M - m)n_c$). Since this trap is considered thermally disconnected, there is no thermal excitation of these electrons in the conduction band. The third equation represents the conservation of charge in the crystal, with the left-hand side equal to the rate of change of the concentration of electrons, and the right-hand side equal to the rate of change of the concentration of holes in the crystal. The last equation expresses the TL intensity that is equal to the rate of change of the concentration of holes in the RC (term $(n + m + n_c)n_c A_h$).

Use the following numerical values:

$$E = 1 \text{ eV}, s = 10^{12} \text{ s}^{-1}, N = 10^{10} \text{ cm}^{-3}, M = 10^{10} \text{ cm}^{-3}, A_n = 10^{-7} \text{ cm}^3 \text{ s}^{-1},$$
$$\beta = 1 \text{ Ks}^{-1}, n_0 = 10^9 \text{ cm}^{-3}, m_0 = 10^9 \text{ cm}^{-3}, A_m = 10^{-5} \text{ cm}^3\text{s}^{-1}, A_h = 10^{-5} \text{ cm}^3\text{s}^{-1}. \text{ A linear heating rate is assumed.}$$

Solution

(a) The following *Mathematica* program integrates the differential equations for the IMTS model. The solution of the system of differential equations (3.16)–(3.18) is stored in the parameter **sol** as in previous exercises, and the command **Plot** is used to graph the functions $n_c(T)$, $n(T)$, $n_h(T)$, $m(T)$, and TL(T).

Since the value of the ratio $A_n/A_h = 0.01$, i.e. the recombination probability coefficient A_h is 100× larger than the retrapping probability coefficient A_n, one would expect that first-order kinetics will dominate the kinetic process: The calculated TL glow curve has indeed the characteristic shape for first-order peaks.

The concentration of electrons in the AT decreases as the temperature increases, while the instantaneous concentration of electrons in the conduction band has a very similar shape to the measured TL glow curve. In addition, the $n_c(T)$ and TL(T) maxima occur at the same temperature, indicating that in this example the TL and TSC peaks are not shifted with respect to each other. The concentration of electrons in the TDDT increases with temperature and reaches saturation after $140°$C, shortly after the end of the TL glow curve.

```
Remove["Global`*"];
E1 = 1.0;  s1 = 10^12;  k1 = 8.617*10^-5;  beta1 = 1;
Npoints = 250;
N1 = 10^10;  M = 10^10;  no = 10^9;  mo = 10^9;  An = 10^-7;
Am = 10^-5;  Ah = 10^-5;
sol = NDSolve[{n1'[x] == -n1[x]*(s1/beta1)*E^(-E1/(k1*(273+x)))
    +An*(N1-n1[x])*nc[x])/beta1, nc'[x] == -n1'[x]-Ah*nc[x]
    *(m[x]+n1[x]+nc[x])/beta1-nc[x]*(M-m[x])*Am/beta1, m'[x]
    == nc[x]*(M-m[x])*Am/beta1, m'[x] == nc[x] * (M-m[x])
    *Am/beta1, n1[0] == no, nc[0] == 0, m[0] == mo},
    {n1, nc, m}, {x, 0, Npoints}, MaxSteps → 50000];
```

```
Plot[Evaluate[nl[x]/.sol], {x, 0, Npoints},
 AxesLabel → {"T", "nl(T)"}, PlotRange → All,
 ImageSize → 753];
Plot[Evaluate[nc[x]/.sol], {x, 0, Npoints},
 AxesLabel → {"T", "nc(T)"}, PlotRange → All,
 ImageSize → 753];
Plot[Evaluate[m[x]/.sol], {x, 0, Npoints},
 AxesLabel → {"T", "m(T)"}, PlotRange → All,
 ImageSize → 753];
Plot[Evaluate[nc[x]*(m[x]+nl[x]+nc[x])*Ah/.sol],
 {x, 0, Npoints}, PlotRange → All,
 AxesLabel → {"T", "TL(T)"}, ImageSize → 753];
```

Exercise 3.6: Analytical Expressions for First-Order Kinetics

As was discussed in Chapter 1, Kitis et al [3] developed the following analytical equation for first-order kinetics peaks. The expression relies on two experimentally measured quantities I_M and T_M, and the third parameter is the activation energy E.

$$I(T) = I_M \exp\left[1 + \frac{E}{kT}\cdot\frac{T-T_M}{T_M} - \frac{T^2}{T_M^2}\cdot\left(1 - \frac{2kT_M}{E}\right)\right.$$
$$\left. \times \exp\left(\frac{E}{kT}\cdot\frac{T-T_M}{T_M}\right) - \frac{2kT_M}{E}\right]. \tag{3.20}$$

In this expression I_M = maximum TL intensity, T_M = temperature at which the maximum intensity occurs and T = temperature in degrees K. It is also noted that the usual kinetic parameters s and n_0 are not present in this analytical expression.

Compare the accuracy of the above expression with the exact solution of the first-order kinetics differential equation, by graphing both glow curves on the same graph.

A precise numerical method of expressing the accuracy of expression (3.20) is by calculating the FOM. The FOM is defined as [3]

$$\text{FOM} = \frac{\sum_p |y_{\text{experimental}} - y_{\text{fit}}|}{\sum_p y_{\text{fit}}}, \tag{3.21}$$

where $y_{\text{experimental}}$ and y_{fit} represent the experimental data and the values of the fitting function, respectively. The summations extend over all the available points.

Calculate the FOM for expression (3.20) for first-order kinetics.

Use the following numerical values: $E = 1.0$ eV, $s = 10^{12}$ s^{-1}, $\beta = 1°C\,s^{-1}$, $n_0 = N = 10^{10}$ cm^{-3}.

Solution

The following *Mathematica* program integrates the first-order kinetics equation (3.10) as in the previous Exercise 3.3, and stores the result of the numerical integration into the parameter **sol**. The third line of the program calculates the maximum value of the TL glow curve and stores it in the parameter **maxTL**. The next line of the program sets up a list named **tlcalc** with the data of the TL glow curve stored as pairs (*T*,TL). Each individual TL intensity is divided by the maximum TL intensity **maxTL**, resulting in a normalized TL glow curve with a maximum TL intensity $I_M = 1$.

The sixth line of the program graphs the calculated normalized TL glow curve. The command **ListPlot** is used to graph the normalized TL glow curve, and the program stores this graph as a graphics object named **gr1**. The next two lines calculate the temperature corresponding to the maximum TL intensity, and stores it in the parameter **tempMax**. The next line calculates the values of the TL glow curve using expression (3.20) with $I_M = 1$, and stores them in the list **tlKitis**.

The next line of the program graphs the list **tlKitis** using the command **ListPlot**, and stores this graph as a graphics object named **gr2**. Finally the program uses the command **Show** to graph both the result of numerically integrating equation (3.10) and the calculated glow curve from equation (3.20) on the same graph.

```
Remove["Global`*"];
E1 = 1.0; s1 = 10^12; k1 = 8.617*10^-5; beta1 = 1;
N1 = 10^10; no = 1*N1;
sol = NDSolve[{n1'[x] == -n1[x]*(s1/beta1)
  *E^(-E1/(k1*(x+273))), n1[0] == no}, {n1}, {x,0,273},
  MaxSteps → 50000];
maxTL = Max[Table[Evaluate[-n1'[x]/.sol], {x, 0, 200}]];
tlcalc = Table[{x+273, First[Evaluate[-n1'[x]/.sol]]/maxTL},
  {x, 0, 200}];
gr1 = ListPlot[tlcalc, PlotJoined → True,
  PlotLabel → "Calculated Normalized TL", ImageSize → 755,
  PlotRange → All];
tempList = Position[Table[Evaluate[-n1'[x]/.sol],
  {x, 0, 200}], maxTL]// Flatten;
tempMax = tempList[[1]]-1+273;
tlKitis = Table[{temp, Exp[1+(E1*(temp-tempMax)/
  (k1*temp*tempMax))-(temp^2*(1-2*K1*tempMax/E1)/tempMax^2)
  *Exp[E1*(temp-tempMax)/(k1*temp*tempMax)]-2*k1*tempMax/
  E1]}, {temp, 273, 473}];
gr2 = ListPlot[tlKitis, ImageSize → 755,
  PlotLabel → "Calculated Kitis TL", PlotRange → All];
Show[{gr1, gr2}, PlotLabel → "Calculated and Kitis TL"];
diffList = tlcalc-tlKitis; diffTL = diffList[[All, 2]];
gr3 = ListPlot[diffTL, PlotRange → All,
  PlotLabel → "Calculated TL minus Kitis TL"];
```

```
sum1 = Apply[Plus, Abs[diffTL]]; Print["sum1 = ", sum1];
Sum2 = Apply[Plus, tlKitis[[All, 2]]];
 Print["sum2 = ", sum2];
Print["FOM = ", sum1/sum2];
```

In the last five lines of the above program, the FOM is calculated by using the two lists, namely **tlcalc** and **tlKitis**. The two lists are subtracted and the result is stored in the list **diffTL**, which is then graphed using the command **ListPlot**. The graph is also stored as the graphic object **gr3**.

The commands **Apply, Abs** and **Plus** are used to calculate the two sums which appear in the expression for the FOM, namely the sums

$$\text{Sum1} = \sum_{p} \left| y_{\text{experimental}} - y_{\text{fit}} \right| \tag{3.22}$$

$$\text{Sum2} = \sum_{p} y_{\text{fit}} \tag{3.23}$$

Finally the program prints out the FOM value as the ratio of sum1 and sum2.

The results of running the above program are shown in Figure 3.15. It is seen that expression (3.20) does an excellent job in approximating the TL glow curve,

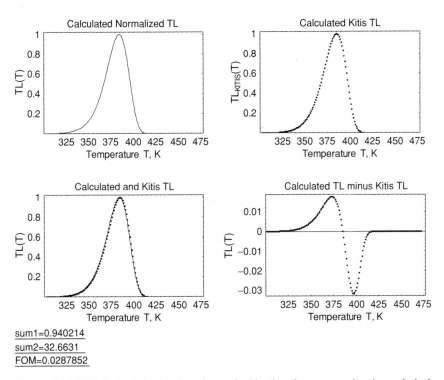

sum1=0.940214
sum2=32.6631
FOM=0.0287852

FIGURE 3.15. FOM calculation for fit to first-order kinetics glow curve using the analytical expression of Kitis et al.

and the two graphs are indistinguishable from each other, at least on the scale of the graph. Furthermore, the differences between the two normalized TL glow curves are between $+0.02$ and -0.03, i.e. between $+2\%$ and -3% of the maximum TL intensity $I_M = 1$. These differences are also known as the **residuals**. The FOM value is equal to 0.0287 which corresponds to a percent accuracy of 2.87% for the goodness of fit.

Exercise 3.7: Analytical Expressions for Second-Order Kinetics

Repeat the previous exercise by using the following second-order kinetics expression developed by Kitis et al [3]:

$$I(T) = 4I_M \, \exp\left(\frac{E}{kT} \cdot \frac{T - T_M}{T_M}\right)$$

$$\times \left[\frac{T^2}{T_M^2} \cdot \left(1 - \frac{2kT}{E}\right)\exp\left(\frac{E}{kT} \cdot \frac{T - T_M}{T_M}\right) + 1 + \frac{2kT_M}{E}\right]^{-2}. \quad (3.24)$$

Compare the accuracy of expression (3.24) with the exact solution of the second-order kinetics differential equation (3.10), by graphing both glow curves on the same graph, and by calculating the FOM.

Use the following numerical values: $E = 1.0$ eV, $s = 10^{12}$ s^{-1}, $\beta = 1$ K s^{-1}, $n_0 = N = 10^{10}$ cm^{-3}.

Solution

The following *Mathematica* program integrates the second-order kinetics equation as in the previous example, and follows exactly the same layout, with the only difference being that it uses expression (3.24) for second-order kinetics.

```
Remove["Global`*"];
E1 = 1.0; s1=10^12; k1 = 8.617*10^-5; beta1 = 1;
 N1 = 10^10; no = 1*N1;
sol = NDSolve[{n1'[x] == -n1[x]^2/N1*(s1/beta1)*E^(-E1/
 (k1*(x+273))), n1[0] == no}, {n1}, {x, 0, 273},
 MaxSteps → 50000];
maxTL=Max[Table[Evaluate[-n1 '[x]/. sol], {x, 0, 200}]];
tlcalc = Table[{x + 273, First[Evaluate[-n1 '[x]/. sol]]/
 maxTL}, {x, 0, 200}];
grl = ListPlot[tlcalc, PlotJoined → True,
 PlotLabel → "Calculated Normalized TL", ImageSize → 755];
tempList = Position[Table[Evaluate[-n1 '[x]/. sol],
 {x, 0, 200}], maxTL]//Flatten;
tempMax = tempList[[1]] - 1 + 273;
```

```
tlKitis = Table[{temp, 4*Exp[E1*(temp - tempMax)/
  (k1*temp*tempMax)]*((temp^2*(1-2*K1*tempMax/E1)/tempMax^2)
  *Exp[E1*(temp-tempMax)/(k1*temp*tempMax)]+1+2*k1*tempMax/
  E1)^-2}, temp, 273, 473];
gr2 = ListPlot[tlKitis, ImageSize → 755,
  PlotLabel → "Calculated Kitis TL"];
Show[{gr1, gr2}, PlotLabel → "Calculated and Kitis TL"];
diffList = tlcalc-tlKitis; diffTL = diffList[[All, 2]];
gr3 = ListPlot[diffTL, PlotRange → All,
  PlotLabel → "Calculated TL minus  Kitis TL"];
sum1 = Apply[Plus, Abs[diffTL]]; Print["sum1=", sum1];
Sum2 = Apply[Plus, tlKitis[[All, 2]]]; Print["sum2=", sum2];
Print["FOM=", sum1/sum2];
```

The results of running the above program are shown in Figure 3.16. The expression of Kitis et al [3] approximates very well the second-order TL glow curve, and the differences between the normalized TL glow curves are between +0.008 and −0.006, i.e. between +0.8% and −0.6% of the maximum height of the TL glow peak. The FOM value is equal to 0.0083, corresponding to a goodness of fit of 0.83%.

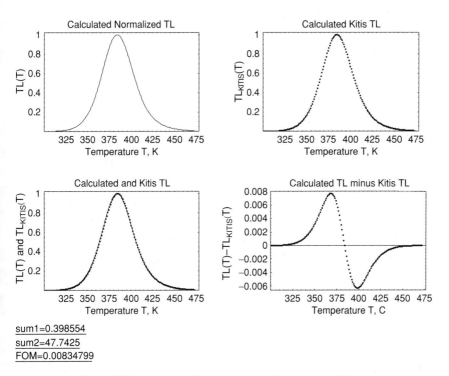

sum1=0.398554
sum2=47.7425
FOM=0.00834799

FIGURE 3.16. FOM calculation for second-order kinetics using *Mathematica*.

Exercise 3.8: Analytical Expressions for General-Order Kinetics

Repeat the previous exercise by using the following general-order kinetics expression developed by Kitis et al [3]:

$$I(T) = I_m \cdot b^{\frac{b}{b-1}} \exp\left(\frac{E}{kT}\frac{T-T_m}{T_m}\right)\left[1 + (b-1)\frac{2kT_m}{E} + (b-1)\left(1-\frac{2kT}{E}\right)\right.$$

$$\left. \times \left(\frac{T^2}{T_m^2}\exp\left(\frac{E}{kT}\cdot\frac{T-T_m}{T_m}\right)\right)\right]^{\frac{-b}{b-1}}. \tag{3.25}$$

Here b = order of kinetics, usually a value between 1 and 2.

The computer program follows the exact same outline as the previous example, with the following line replacing the corresponding line for second-order kinetics.

```
tlKitis=
Table[{temp, (b^(b/(b-1)))*Exp[E1*(temp-tempMax)/
(k1*temp*tempMax)]*((b-1)*(temp^2*(1-2*k1*tempMax/E1)/
tempMax^2)*Exp[E1*(temp-tempMax)/(k1*temp*tempMax)]+1
+(b-1)*2*k1*tempMax/E1)^(-b/(b-1))}, {temp, 273, 723}];
```

The FOM value is equal to 0.0091 which corresponds to a percent accuracy of 0.91% for the goodness of fit (Figure 3.17).

Exercise 3.9: Comparative Study of the Accuracy of Analytical Expressions for First-Order TL Glow Peaks

The analytical expressions A–F listed below are found in the literature, and they all describe a single first-order glow peak with kinetic parameters E and s.

(a) Calculate a synthetic reference glow peak (RGP) of first-order kinetics with trapping parameters $E = 1$ eV and $s = 10^{12}$ s^{-1}, $n_0 = 10^6$ cm^{-3}.
(b) Investigate the accuracy of the following expressions A–F by calculating the FOM for each expression by using the RGP from part (a).

$$A\,[5,6]: \quad I(T) = As\exp\left(-\frac{E}{kT}\right)\exp\left[-\frac{skT^2}{\beta E}\exp\left(-\frac{E}{kT}\right)\left(1-\frac{2kT}{E}\right)\right]. \tag{3.26}$$

Expression (3.26) can be easily derived from the Randall–Wilkins equation (1.5) for first-order kinetics by replacing the parameter n_0 with the Area A under the glow curve, and by using the series approximation of the integral in the first-order

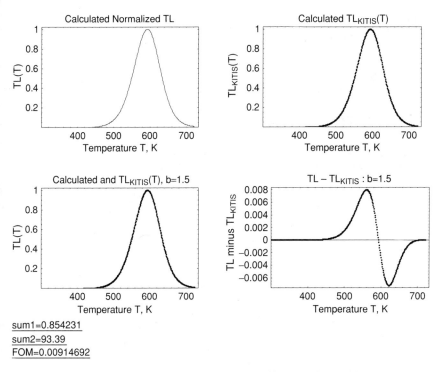

sum1=0.854231
sum2=93.39
FOM=0.00914692

FIGURE 3.17. FOM calculation for general-order kinetics using *Mathematica*.

kinetics, equation (1.52).

$$B\ [5,6]:\ I(T) = I_M \exp\left\{1 + \frac{E}{kT_M}\left(\frac{T - T_M}{T_M}\right) - \exp\left[\frac{E}{kT_M}\left(\frac{T - T_M}{T_M}\right)\right]\right\}$$

(3.27)

$$C\ [5,6]:\ I(T) = I_M \exp\left\{1 + \frac{E}{kT}\left(\frac{T - T_M}{T_M}\right) - \exp\left[\frac{E}{kT}\left(\frac{T - T_M}{T_M}\right)\right]\right\}.$$

(3.28)

Expression (3.28) is derived by approximating a linear heating rate function by a hyperbolic heating rate profile. It is sometimes referred to as the Hyperbolic approximation.

$$D\ [7]:\ I(T) = I_M \exp\left\{1 + \frac{E}{kT}\left(\frac{T - T_M}{T_M}\right)\right.$$
$$\left. - \frac{T^2}{T_M^2}\exp\left[\frac{E}{kT}\left(\frac{T - T_M}{T_M}\right)\right](1 - \Delta_M) - \Delta_M\right\},$$

(3.29)

with, $\Delta = 2kT/E$ and $\Delta_M = 2kT_M/E$.

Equation (3.29) is the first-order equation developed by Kitis et al [3], referred to several times in this book.

$$\text{E [5, 6]: } I(T) = \frac{A}{\sqrt{2\pi}\,[\sigma - \alpha(T - c)]}\exp\left\{-\frac{(T - c)^2}{2[\sigma - \alpha(T - c)^2]}\right\}. \quad (3.30)$$

Expression (3.30) is termed as a "modified Gaussian" curve and relies on a total of four adjustable parameters A, α, σ, and c while the rest of the functions in this exercise are based on three adjustable parameters. The activation energy E is deduced from the width of the modified Gaussian.

$$\text{F [8]: } I(T) = 2.713 I_M\left(\frac{T - T_M}{b} + 0.996\right)^{15}\exp\left\{-\frac{T - T_M}{b} + 0.996\right\}^{16}. \quad (3.31)$$

Expression 3.31 is known as the Weibull approximation to the first-order glow peak, and the activation energy is given by the expression

$$E = T_M\frac{k}{b}[-b + \sqrt{7b^2 + 242.036 T_M^2}], \quad (3.32)$$

where b = width of the Weibull function.

In the above expressions A = area under the glow curve (counts), E = activation energy (eV), s = frequency factor (s^{-1}), T = temperature (K), k = Boltzmann constant (eV K^{-1}), I_M = maximum intensity of the glow peak (counts per K), T_M = temperature at peak maximum (K).

Solution

The glow curves of TL materials are in most cases complex curves consisting of many overlapping glow peaks. The deconvolution of complex glow curves into their individual glow peaks (glow-curve-deconvolution—GCD) is widely applied for dosimetric purposes and for evaluating the trapping parameters E and s using curve fitting methods.

The purpose of this exercise is to investigate the accuracy of the expressions A–F from the literature by calculating the FOM for each of these analytical expressions and for the given RGP.

The success of each one of the above expressions in fitting the RGP can be tested by two possible procedures. The first and simplest possibility is to evaluate the parameters I_M and T_M from the RGP and to insert them directly in each one of the listed first-order kinetic expressions, in order to obtain the respective peak. The glow peak obtained in this manner is then directly compared with the RGP through the FOM value.

TABLE 3.4. FOM values for a variety of analytical expressions for first-order kinetics

Peak	Integral	I_M	E	T_M	S	FOM(I)%	FOM(II)%
RGP	999261.5	30664.5	1	384.6	10^{12}		
A	999234.1	30662.0	1.00012	384.5	1.01×10^{12}	0.013	0.49
B	993623.0	31154.0	1.09	385.66		4.84	5.8
C	999124.0	30649.5	1.027	384.37		0.8	2.6
D	999260.1	30663.9	1.00018	384.494		0.0058	0.019
E	998951.4					1.1	–
F	1000098.0	30700.0	1.001	384.54		0.027	0.23

The second possible procedure is to use a computerized glow-curve fitting procedure by using commercially available software. In this exercise the curve fitting procedure was performed using the MINUIT program [9].

The results are shown in Table 3.4. The column labeled FOM(I) is obtained by using the MINUIT program where E, I_M, and T_M are free parameters. The parameters α, σ, and c in expression E are also treated as adjustable fitting parameters. The column labeled FOM(II) is obtained when the values of E, I_M, and T_M corresponding to the RGP are inserted into the expressions A–F and the respective glow peak is evaluated.

It is noted that the analytical expression E (the "modified Gaussian" curve), relies on several adjustable fitting parameters (α, c, σ) for which there is no analytical expression in the literature. As a consequence, only results for FOM(II) are given in Table 3.4. For the rest of the analytical expressions, both FOM(I) and FOM(II) values are listed in Table 3.4.

Both the FOM(I) and FOM(II) values give an estimate of the accuracy by which each expression fits the RGP. It is important to note that the FOM(II) values are very sensitive to small changes in the parameters IM,TM used in expressions A–F in this exercise.

The final conclusions for each expression are summarized as following:

(i)   The expression which best fits the RGP is expression D by Kitis et al.
(ii) Although expression A is mathematically equivalent to expression D, it gives a slightly higher FOM. The reason is that in expression A the frequency factor is a free parameter and it is not easy to achieve very small variation steps for the parameters during the curve fitting procedure.
(iii) Expressions B and C give very poor FOM values.
(iv) On the other hand, expressions E and F although not physically based, seem to give a better fit to the RGP than expressions B and C.
(v)   Expression D is clearly preferable from the rest of the functions in the list for GCD analysis of complex TL glow curves.
(vi) On the other hand, expression F is included in several commercially available software packages (such as SigmaPlot or PeakFit), and can be used as an accurate first-order glow peak algorithm.

(vii) If only the glow-peak area is of interest, then all expressions provide a good estimate of it as seen in column 2 of the Table 3.4, with an accuracy better than 1% from the reference value of $A = n_0 = 10^6$.

Exercise 3.10: Comparative Study of the Accuracy of Analytical Expressions for General-Order TL Glow Peaks

The analytical expressions A–E listed below are found in the literature, and all describe a single general-order glow peak with kinetic parameters E, b, and s.

(a) Calculate a synthetic RGP for second-order kinetics with trapping parameters $E = 1$ eV, $s = 10^{12}$ s^{-1}, $n_0 = N = 10^6$ cm^{-3}.

(b) Use the second-order RGP from part (a) to investigate the accuracy of the following expressions A–E. Calculate the FOM for each expression.

(c) The general-order expressions in the list below cannot be used for $b = 1$ because of the presence of the ratio $b/(b-1)$. Nevertheless, these expressions can still be used to fit a first-order glow peak by using a value of b very close to 1. Use the general-order expression C and find the value of the kinetic order b for which this general order expression fits best a first-order RGP.

(d) Having found the appropriate value of b in part (c), estimate the accuracy with which the expressions A, B, and D fit the first-order RGP.

$$A\,[5,6]: I(T) = As\,\exp\left(-\frac{E}{kT}\right)$$

$$\times\left[1 + (b-1)\frac{skT^2}{\beta E}\exp\left(-\frac{E}{kT}\right)\left(1 - \frac{2kT}{E}\right)\right]^{-\frac{b}{b-1}} \quad (3.33)$$

$$B\,[7]: I(T) = I_M\,b^{\frac{b}{b-1}}\exp\left(\frac{E}{kT}\frac{T-T_M}{T_M}\right)$$

$$\times\left[(b-1)(1-\Delta)\frac{T^2}{T_M^2}\exp\left(\frac{E}{kT}\frac{T-T_M}{T_M}\right) + Z_M\right]^{-\frac{b}{b-1}}, \quad (3.34)$$

where $Z_M = 1 + (b-1)\Delta_M$

$$C\,[10]: I(T) = I_M\,\exp\left(\frac{E}{kT}\frac{T-T_M}{T_M}\right)\left[1 + \frac{b-1}{b}\frac{E}{kT_M}\right.$$

$$\times\left.\left\{\frac{T}{T_M}\exp\left(\frac{E}{kT}\frac{T-T_M}{T_M}\right)F\left(\frac{E}{kT}\right) - F\left(\frac{E}{kT_M}\right)\right\}\right]^{-\frac{b}{b-1}},$$

$$(3.35)$$

where $F(x)$ is defined by

$$F(x) = 1 - \frac{a_0 + a_1 x + x^2}{b_0 + b_1 x + x^2}, \quad (3.36)$$

with

$$\alpha_0 = 0.250621 \quad b_0 = 1.681534$$

$$\alpha_1 = 2.334733 \quad b_1 = 2.330657$$

D [10]: $I(T) = I_M \exp\left(\dfrac{E}{kT_M^2}(T - T_M)\right)$

$$\times \left[\dfrac{1}{b} + \left(\dfrac{b-1}{b}\right) \exp\left(\dfrac{E}{kT_M^2}(T - T_M)\right)\right]^{-\frac{b}{b-1}} \tag{3.37}$$

E [11]: $I(T) = 5.2973\, I_M \left[1 + \exp\left\{-\left(\dfrac{T - T_M}{a_2} + 0.38542\right)\right\}\right]^{-2.4702}$

$$\times \exp\left\{-\left(\dfrac{T - T_M}{a_2} + 0.38542\right)\right\}. \tag{3.38}$$

Equation (3.38) is a special case of the four-parameter logistic asymmetric function, which was shown to describe accurately second-order TL glow peaks [11]. The activation energy in the case of expression E is given by

$$E = kT_M \left[-2 + \sqrt{4 + 1.189\dfrac{T_M^2}{a_2^2}}\,\right], \tag{3.39}$$

where a_2 represents the width of the logistic asymmetric function.

The rest of the symbols in the above expressions are A = area under the glow peak (counts), E = activation energy (eV), s = frequency factor (s^{-1}), b = kinetic order, T = temperature (K), k = Boltzmann constant (eV K^{-1}), I_M = Intensity at peak maximum (counts K^{-1}), T_M = temperature at peak maximum (K).

Solution

(a) The purpose of this exercise is to investigate the accuracy of the expressions A–E from the literature by calculating the FOM for each of these analytical expressions and for the given RGP.

The RGP necessary to test the expressions can easily be obtained by numerical integration of the exponential integral (see exercises 3.3 and 3.7 in Chapter 3). As an alternative to numerical integration, Expression A is suggested for the creation of the RGP.

The success of each one of the above expressions to fit the RGP can be tested by the two procedures described in the previous exercise. The results are shown in Table 3.5. The values of FOM(I) are obtained from the curve fitting procedure where E, I_M, and T_M are free parameters, whereas FOM(II) is obtained when the values of E, I_M, and T_M corresponding to the RGP are inserted into the expressions A–E and the respective glow peak is evaluated.

Both FOM(I) and FOM(II) values are seen to give the same order of magnitude of FOM values. The final conclusions for each expression are summarized as

TABLE 3.5. FOM values for a variety of analytical expressions for general-order kinetics

Peak	Integral	I_M	E	T_M	s	FOM(I)%	FOM(II)%
RGP	996733.6	20876.1	1	383.9	10^{12}		
A	996728.9	20875	1.00009	384	1.009×10^{12}	0.017	0.66
B	996734.7	20874.9	1.00009	383.73		0.0016	0.7
C	996734	20876.1	0.99998	383.84		0.00003	0.2
D	990456.3	20954.6	1.096	385.36		4.8	6.6
E	1004142	20870	0.969	383.62		1.4	–

following:

(i) Expressions A, B, and C result in almost perfect fits for the second-order RGP.
(ii) Expression D gives very poor results.
(iii) Expression E, although not based on a physically meaningful model fits better the RGP than expression D.
(iv) On the other hand, expression E is included in several commercially available software packages and can be used as a second-order glow peak algorithm.
(v) The difference between FOM(I) and FOM(II) for expressions A, B, and C is very high although the difference between the values of E, I_M and T_M of the RGP is negligible compared with those obtained by the curve fitting procedure.
(vi) In the case of expression D the difference between FOM(I) and FOM(II) is very low and the differences between the values of E, I_M, and T_M of the RGP are negligible compared with those obtained by the curve fitting procedure.
(vii) If only the glow-peak area is of interest, then all expressions A–E give an accurate estimate of it, with an accuracy better than 1%.

(b) The general-order equation is not valid for $b = 1$ because of the presence of the term $b/(b-1)$ in expressions A–D. However, it could be valid for b slightly higher than 1. To investigate whether the general-order expressions in this exercise can be used to fit a first-order glow peak, we will use expression C. The first-order kinetics RGP is obtained as in the previous example with the parameters $E = 1$ eV and $s = 10^{12}$ s^{-1}.

The purpose of this investigation is not a simple comparison of various expressions, but to examine the possibility of using a general-order expression to fit a first-order glow peak. This in turn simplifies the development of a single algorithm that can be used for all values of the kinetic order b.

Using the MINUIT program [9], the curve fitting procedure was applied for the b values listed in column 1 of Table 3.6. The FOM(I) values obtained are listed in column 2.

It is clear that a general-order expression fits very accurately with the first order RGP for $b < 1.05$. However, the best b value is 1.000005.

TABLE 3.6. FOM values for
several values of b for first-order
kinetics

b	FOM(I)%
1.05	1.8
1.005	0.18
1.0005	0.017
1.00005	0.0011
1.000005	0.0007
1.0000005	0.00073

(c) Having found the value of $b = 1.000005$ for which the general-order expression C best approximates the first-order kinetics RGP, the expressions A, B, and D were tested by inserting the values of I_M, T_M, and E of the RGP into expressions A, B, and D and calculating the FOM values. The results are shown in Table 3.7.

The results in Table 3.7 show that the general-order kinetics expressions A, B, and C approximate very well the first-order kinetics RGP when a value of $b = 1.000005$ is used.

Exercise 3.11: Numerical Study of Mixed-Order Kinetics

Certain TL glow peaks have been known to exhibit "mixed order" kinetics. The mixed order kinetic equation is given by the expression [12, 13]:

$$I(T) = h^2 s' \alpha \exp\left(-\frac{E}{kT}\right) \frac{\exp\left(hs'\frac{kT^2}{\beta E}\exp\left(-\frac{E}{kT}\right)\left(1 - \frac{2kT}{E}\right)\right)}{\left[\exp\left(hs'\frac{kT^2}{\beta E}\exp\left(-\frac{E}{kT}\right)\left(1 - \frac{2kT}{E}\right)\right) - \alpha\right]^2},$$

(3.40)

where in some models the pre-exponential factor is given by $s' = s/N$, while in other models it is given by $s' = s/(N + h)$.

TABLE 3.7. FOM values for several
values of b for a variety of
analytical expressions

Expression	FOM(I)%
A	0.0058
B	0.0059
C	3.76

The parameter $\alpha = n_0/(n_0 + h)$ is the parameter characterizing the mixed-order kinetics. As $\alpha \to 0$ equation (3.40) reduces to the expression for first-order kinetics, and as $\alpha \to 1$ it reduces to second-order kinetics.

The purpose of this exercise is to perform a systematic study of the behavior of the glow peaks following mixed-order kinetics using the parameters $E = 1$ eV, $s = 10^{12}$ s^{-1}, and $n_0 = N = 10^3$ m^{-3}, and assuming two different cases, case I with $s' = s/(N + h)$ and case II with $s' = s/N$.

For each of the two cases I and II calculate the following and compare the results.

(a) Study the shape of mixed-order glow peaks as a function of the parameter α.
(b) Compare the behavior of the parameter α and of the kinetic order b as a function of symmetry factor μ_g.
(c) Evaluate the behavior of the peak maximum temperature T_M as a function of α and compare it with the corresponding behavior of a general-order kinetics glow peak.
(d) Repeat (b) for the full width at half maximum (FWHM $= \omega$)
(e) Evaluate the activation energy of the mixed-order glow peaks using the peak shape equations (1.48) for general-order glow peaks and discuss the validity of these equations for mixed-order kinetics.
(f) Find a graphical relation between the mixed-order kinetics parameter α and the general-order kinetics parameter b.
(g) Compare the shape of mixed and general-order glow peaks of the same symmetry factor.

Solution

All calculations needed for the present exercise can be easily performed in a spreadsheet program or in *Mathematica*. Care must be taken concerning the numerical accuracy of the evaluations which will depend upon the temperature increment used in the glow-peak evaluation. For example, if one evaluates the TL intensity every 1 K ($\Delta T = 1$ K) the error in the resulting values of T_M and FWHM will be of the order of 0.5 K. In the evaluations below the *Mathematica* program was used with a very small temperature interval $\Delta T = 0.001$ K, since there was no substantial difference in the calculation time if a value of $\Delta T = 1$ K was chosen instead. Equation (1.7) was used to calculate the general-order kinetics glow peaks.

Case I: $s' = s/(N + h)$

Taking into account that $s' = s/(N + h)$ and $n_0 = N$ we can find after some algebra that

$$h^2 s' = \frac{(1 - \alpha)^2}{\alpha} s n_0 \tag{3.41}$$

and

$$hs' = (1 - \alpha)s. \tag{3.42}$$

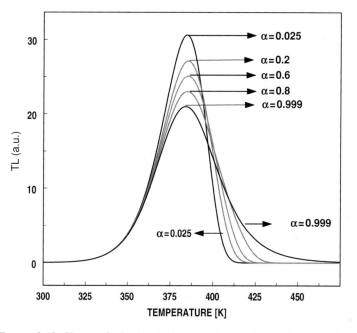

FIGURE 3.18. Shapes of mixed-order kinetics glow peaks as a function of α.

By inserting equations (3.41) and (3.42) in (3.40) we can proceed to the required calculations by noting that $I(T)$ depends only on the known parameters E, α, s, and n_0.

(a) The shapes of mixed-order glow peaks as a function of the parameter α are shown in Figure 3.18. The glow-peak shapes behave in a manner similar to that of the general-order kinetics (see Figure 1.6).

(b) Figure 3.19 shows the behavior of the parameter α and of the kinetic order b as a function of the symmetry factor μ_g.

The parameter α is given in values of $(\alpha + 1)$ in order to fit on the same scale with the kinetic order b. From Figure 3.19 we can see that the mixed- and general-order glow peaks coincide exactly when $\alpha + 1 = b$. This happens for first-order kinetics where $\alpha \to 0$, and for values of the symmetry factor $\mu_g \geq 0.50$. In the intermediate cases the mixed- and general-order glow peaks show some differences, even in cases where they have the same symmetry factor (see Figure 3.24 below).

(c) Figure 3.20 shows the glow peak maximum temperature T_M as a function of α and b.

Note that the y-axis covers a temperature region of only 2 K. The behavior of glow peaks with mixed-order kinetics is very different from that of general-order peaks, although the peak maximum variation is less than 1.5 K for values of b between 1 and 2, and for values of α between 0 and 1. These differences are seen here because the glow peaks were evaluated using a temperature increment of $\Delta T = 0.001$ K and therefore the accuracy of T_M was better than 0.001 K.

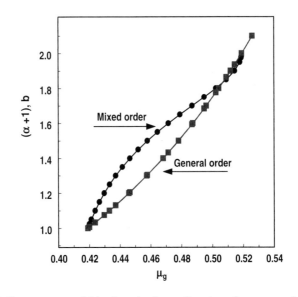

FIGURE 3.19. Parameter α and kinetic order b as a function of symmetry factor. The parameters α is used as $\alpha + 1$ in order to fit on the same scale with b.

(d) Figure 3.21 shows the variation of the FWHM $= \omega$ with α and b. The variation of the FWHM as a function of α in the case of mixed-order glow peaks is very similar to the corresponding behavior of the general-order glow peaks. The slight differences are seen here because the accuracy is of the order of 0.001 K.

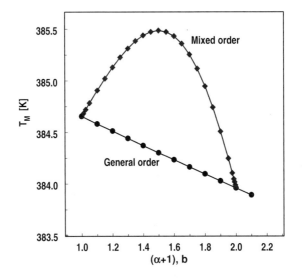

FIGURE 3.20. Peak maximum temperature as a function of α and b. The parameter $\alpha + 1$ is used instead of α, in order to fit on the same scale as b. Note the extent of the y-scale.

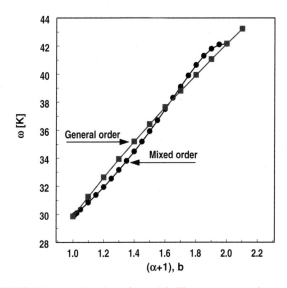

FIGURE 3.21. FWHM (ω) as a function of α and b. The parameter α is used as ($\alpha + 1$) in order to fit on the same scale as b.

In the case of experimental measurements, this difference will not be easily seen.

(e) Figure 3.22 shows the activation energy E as a function of α.

The peaks shape equations (1.48) are used to calculate the energy E [13]. The results of Figure 3.22 show that when the peak shape methods are applied to mixed-order glow peaks, they give the correct values of E for $\alpha \to 0$ and $\alpha > 0.8$. In the intermediate cases the methods underestimate E by less than 2% for the methods based on τ and δ, and less than 3–4% for the methods based on ω.

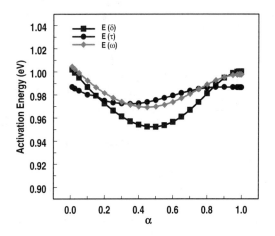

FIGURE 3.22. Activation energies E as a function of α calculated using equation (1.48).

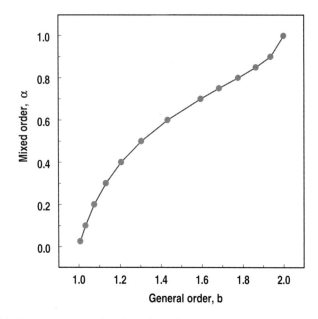

FIGURE 3.23. Parameter α as a function of the kinetic order b, for glow peaks having the same symmetry factor.

These results are in agreement with those of Chen et al [12]. Therefore, we conclude that the general peak shape equations (1.48) can be applied in the case of mixed-order kinetics without any significant loss in accuracy.

(f) Graphical relation of the parameter α as a function of the kinetic order b.

This relationship can be found by using trial and error, and by identifying mixed- and general-order glow peaks having the same symmetry factor μ. By requiring that the values of the symmetry factor differ by less than 10^{-4}, the appropriate values of α and b were found and plotted in Figure 3.23.

(g) Comparison of mixed- and general-order glow peaks having the same symmetry factor.

As was discussed above (Figure 3.19), differences between mixed-order and general-order glow peaks exist only in the intermediate values of α and b. Figure 3.24 shows a comparison between mixed- and general-order glow peaks having the same symmetry factor. The general-order glow peak (solid line) corresponds to $b = 1.4325$ and a symmetry factor 0.471502, whereas the mixed-order glow peak (dash line) correspond to $\alpha = 0.6$ and a symmetry factor 0.471503.

Case II: $s' = s/N$

If we now consider the value of $s' = s/N$ and take $n_0 = N$, we find after some algebra that

$$h^2 s' = \left(\frac{1 - \alpha}{\alpha} \right)^2 s n_0 \qquad (3.43)$$

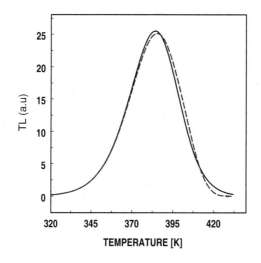

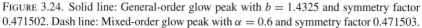

FIGURE 3.24. Solid line: General-order glow peak with $b = 1.4325$ and symmetry factor 0.471502. Dash line: Mixed-order glow peak with $\alpha = 0.6$ and symmetry factor 0.471503.

and

$$hs' = \frac{1 - \alpha}{\alpha} s \tag{3.44}$$

Therefore the behavior of the glow peaks in case II can be simulated by using equation (3.40) with the values of $h^2 s'$ and hs' given by equations (3.43) and (3.44).

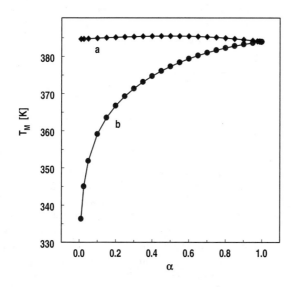

FIGURE 3.25. Behavior of TM as a function of the parameters α. Curve (a) is for Case I and curve (b) is for Case II.

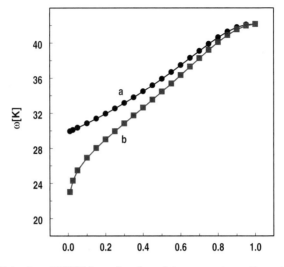

FIGURE 3.26. Behavior of FWHM as a function of the parameter α. Curve (a) is for Case I and curve (b) is for Case II.

The results obtained are the same as those found in Case I with accuracy better that 10^{-3}. The only substantial differences were found in the behavior of T_M and FWHM.

The different behavior of T_M is shown in Figure 3.25. The reason for this difference between Cases I and II is that the pre-exponential factor in Case I varies according to $1 - \alpha$, whereas in Case II it varies according to $(1 - \alpha)/\alpha$.

Therefore, equation (3.40) in Case II has an additional dependence on $1/\alpha$. This factor has a functional behavior similar to that of the heating rate $1/\beta$, causing T_M to increase as the value of α is increased.

Differences are also seen in the behavior of FWHM, which are shown in Figure 3.26.

References

[1]   H. Gould and J. Tobochnik 1988 *An Introduction to Computer simulation methods* (New York: Addison-Wesley Publishing Co)
[2]   C.M. Sunta, W.E.F. Ayta, J.F.D. Chubaci, S. Watanabe: Rad. Meas. 35, 47 (2002).
[3]   G. Kitis, J.M. Gomez-Ros, J.W.N. Tuyn, J. Phys. D 31 (1998) 2636.
[4]   C.M. Sunta, W.E.F. Ayta, T.M. Piters, S. Watanabe. Rad. Meas. 30, 197 (1999).
[5]   A.J.J. Bos, T.M. Piters, J.M. Gomez Ros and A. Delgado. Radiat. Prot. Dosim. 47 (1993) 473-477 and Radiat. Prot. Dosim. 51(1994)257-264.
[6]   Y.S. Horowitz and D. Yossian, Radiat, Prot. Dosim. 60(1) 1995.
[7]   G. Kitis, J. M. Gomez-Ros and J.W.N. Tuyn J. Phys. D: Appl. Phys. 31 (1998) 2636.
[8]   V. Pagonis, S.M. Mian and G. Kitis Radiat. Prot. Dosimetry 93 (2001) 11-17

[9]   F. James and M. Roos, Minuit, CERN program Library, entry D506, http://consult.cern.ch/writeups/minuit.

[10]   J.M. Gomez-Ros and G. Kitis Radiat. Prot. Dosimetry 101 (2002) 47-52

[11]   V. Pagonis and G. Kitis Radiat. Prot. Dosimetry 95 (2001) 225-229

[12]   R. Chen, N. Kristianpoller, Z. Davidson and R. Visocekas. Mixed first and second-order kinetics in thermally stimulated processes. Journal of Luminescence 23 (1981) 293-303.

[13]   R. Chen and S.W.S. McKeever. Theory of Thermoluminescence and Related Phenomena, World Scientific 1997

4
TL Dose Response Models

Introduction

The dependence of the Thermoluminescence (TL) signal on the dose imparted on TL materials is of great practical importance in the fields of radiation protection dosimetry and in dating applications. In this chapter several examples of theoretical models are given that have been used to explain the TL versus dose response of a variety of materials.

The first section of this chapter gives an overview of nonlinear response exhibited by several materials, and introduces the terminology used in describing these nonlinearities. The rest of the chapter gives specific exercises based on published kinetic models in the TL literature. The first exercise is based on the one-trap one-recombination center model and demonstrates several basic characteristics of such models. In particular, this model shows the importance of incorporating appropriate relaxation periods after each irradiation or heating stage in the simulation. The second exercise is typical of a class of kinetic models based on the existence of competing traps, and shows how competition between traps during irradiation can lead to superlinear behavior in the TL dose response of materials. The third exercise describes a more complex case in which competition phenomena are of importance during *both* the excitation and the heating stage of TL.

The fourth exercise in this chapter provides an example of obtaining the superlinearity index $g(D)$ and the supralinearity index $f(D)$ from experimental TL versus dose curves. The quantities $f(D)$ and $g(D)$ are defined in the next section.

The computer programs are written in *Mathematica*, and the results of the programs are compared with specific numerical data in the original published papers. Special effort was made to use the same notation as in the original papers, in order to facilitate cross-referencing in the literature for readers interested in pursuing these models further.

The programs in this chapter have a similar structure consisting of a main procedure which calls several subroutines to solve the systems of differential equations and to reinitialize the parameters between the irradiation and relaxation stages in the simulations. It is our hope that this "modular" programming will make it easier for interested readers to modify the programs and to develop their own code for a variety of purposes.

Overview of Nonlinear Dose Response of TL Materials and Terminology

In this section some fundamental concepts and terminology is presented, relevant to experimentally observed nonlinear dose response of TL materials to radiation. A comprehensive review of the various theoretical aspects of the models presented in this section can be found, for example, in the book by Chen and McKeever [1], and in the review article by McKeever and Chen [2].

Several important TL materials exhibit a nonlinear dose response that can be expressed in the mathematical form:

$$I_{\text{max}} = aD^k \tag{4.1}$$

where I_{max} represents the maximum TL intensity (or the TL integral), D is the dose of the radiation and a, k are constants. When I_{max} is plotted as a function of the dose D on a log–log scale, this equation yields a straight line with a slope k that may be larger than unity. For example, Halperin and Chen [3] found that the dose response of UV irradiated diamonds exhibited a slope k between 2 and 3 at certain wavelengths. These authors suggested the term *superlinearity* to describe this more than linear dose response.

Chen and McKeever [4] suggested using the term *superlinearity* for the increase of the derivative of the dose of the TL response function. If the measured TL signal is $S(D)$, the increase in the derivative of $S(D)$ is expressed by the fact that the second derivative $d^2S/dD^2 > 0$. Cases where $d^2S/dD^2 < 0$ are characterized as *sublinear*, and cases where $d^2S/dD^2 = 0$ are characterized as a *linear* dose response. These authors defined the following dimensionless quantity termed the *superlinearity index* $g(D)$:

$$g(D) = \left[\frac{DS''(D)}{S'(D)} \right] + 1. \tag{4.2}$$

As long as $S'(D) > 0$, a value of $g(D) > 1$ signifies *superlinearity*, while a value of $g(D) = 1$ denotes a *linear* dose response, and $g(D) < 1$ signifies *sublinearity*. In the special case where $S(D) = \alpha D^k + \beta$, one obtains $g(D) = k$. The application of equation (4.2) requires the knowledge of an analytical expression which can fit the experimentally obtained TL dose response curves. Otherwise, this equation cannot be applied to the experimental data.

In some low dose ranges several TL materials show a linear dose dependence, followed by a superlinear dose range and by a sublinear range while approaching saturation. For example, gamma-irradiated LiF is known to exhibit such a behavior, as reported for example by Cameron et al [5]. This type of linear-superlinear-saturation behavior can be explained within the framework of TL models based on competition between traps during the excitation of the sample. A typical example of a TL model based on *competition during excitation* is given in Exercise 4.2.

Chen and McKeever suggested using the term *supralinearity* to describe this particular nonlinear dose response. Some authors [6] quantified this behavior by introducing the following dimensionless function termed the *supralinearity index* or *dose response function* $f(D)$:

$$f(D) = [S(D)/D] / [S(D_1)/D_1] \qquad (4.3)$$

where D_1 is a normalization dose in the initial linear range. Values of $f(D) > 1$ indicate values of $S(D)$ above the initial linear range.

In summary, superlinearity is a measure of the rate of change of the dose response with the dose, and is described quantitatively by the *superlinearity index* $g(D)$. On the other hand, supralinearity is described quantitatively by the *supralinearity index* $f(D)$, and is mostly used in TL applications in dating and dosimetry.

Rodine and Land [7] suggested that the superlinear behavior of some materials might be explained by *competition during the heating phase*, instead of competition during the excitation stage. Within the framework of such models, the initial dose dependence of the TL response may be quadratic in nature. Kristiapoller et al [8] developed a mathematical formulation of TL models of this type.

Chen and Fogel [9] discussed some of the disadvantages of these two separate approaches, especially the main assumption that when competition occurs during the irradiation stage, no competition takes place during the heating phase, and vice versa. These authors developed a model that combines the characteristics of the two models: competition during heating *and* competition during excitation models. A typical example of the TL model of Chen and Fogel [9] based on both competition approaches is given in Exercise 4.3.

Exercise 4.1: The Filling of Traps in Crystals During Irradiation

Write a computer program to integrate the kinetic rate equations relevant to the TL model shown in Figure 4.1 [10]. The model consists of one trapping state and one recombination center. The arrows in Figure 4.1 indicate the allowed electron and hole transitions from the conduction and the valence band.

The trap is characterized by total concentration N in the crystal and by instantaneous electron occupancy $n(t)$. The recombination center has instantaneous hole occupancy $n_h(t)$ and total concentration N_h in the crystal. The functions $n_c(t)$ and $n_v(t)$ represent the instantaneous concentrations of free electrons in the conduction band and free holes in the valence band correspondingly. The equations describing the rate of change of the functions $n(t), n_h(t), n_c(t),$ and $n_v(t)$ during the irradiation process are [10]

$$\frac{dn}{dt} = n_c(N - n)A \qquad (4.4)$$

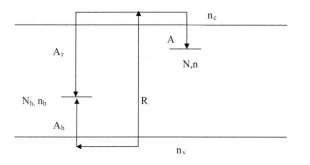

FIGURE 4.1. Kinetic model for the filling of traps during crystal irradiation.

$$\frac{dn_v}{dt} = R - n_v(N_h - n_h)A_h \tag{4.5}$$

$$\frac{dn_h}{dt} = n_v(N_h - n_h)A_h - n_c n_h A_r \tag{4.6}$$

$$\frac{dn_c}{dt} + \frac{dn}{dt} = \frac{dn_h}{dt} + \frac{dn_v}{dt}. \tag{4.7}$$

The first equation expresses mathematically the fact that electrons in the conduction band can be trapped into the electron trap. The second equation describes the process by which free holes in the valence band are created at a constant rate R during the excitation, and these holes can also be trapped from the valence band into the recombination center as indicated by the term $-n_v(N_h - n_h)A_h$. The third equation expresses the fact that the concentration of holes in the recombination center is changed by either trapping electrons from the conduction band (term $-n_c n_h A_r$), or by trapping holes from the valence band (term $n_v(N_h - n_h)A_h$). The last equation (4.7) expresses the conservation of total charge in the crystal, with the left-hand side being equal to the total instantaneous concentration of electrons, and the right-hand side representing the total concentration of holes in the crystal at any time t.

The parameters in the above expressions are as follows:

A = transition probability coefficient of electrons into the trap ($cm^3\ s^{-1}$)

A_h = trapping probability coefficient of holes from the valence band into the recombination center ($cm^3\ s^{-1}$)

A_r = recombination probability coefficient of electrons from the conduction band into the recombination center ($cm^3\ s^{-1}$)

n = instantaneous concentration of electrons in the electron trap at time t (cm^{-3})

N = total concentration of electron traps in the crystal (cm^{-3})

$(N - n)$ = instantaneous concentration of empty main traps available at time t

n_h = instantaneous concentration of holes in the recombination center (cm^{-3})

N_h = total concentration of holes in the crystal (cm^{-3})

n_c = instantaneous concentration of electrons in the conduction band (cm^{-3})

n_v = instantaneous concentration of holes in the valence band (cm^{-3})

R = constant rate of production of electron–hole pairs per cm^3 per second $(\text{cm}^{-3}\,\text{s}^{-1})$

Use the following numerical values $N = 10^{15}\,\text{cm}^{-3}$, $N_h = 3 \times 10^{14}\,\text{cm}^{-3}$, $A = 10^{-17}\,\text{cm}^3\,\text{s}^{-1}$, and $A_r = 10^{-13}\,\text{cm}^3\,\text{s}^{-1}$, $A_h = 10^{-15}\,\text{cm}^3\,\text{s}^{-1}$, $R = 10^{14}\,\text{cm}^{-3}\,\text{s}^{-1}$. The initial conditions at time $t = 0$ are $n(0) = n_h(0) = n_c(0) = n_v(0) = 0$.

(a) Obtain and graph the solution of these coupled differential equations by assuming that the sample is irradiated for time t. By varying the irradiation time t, obtain and graph the functions $n(t)$ at the end of the irradiation period as a function of time t. Show that this yields a nonlinear function $n(t)$ for the filling of the traps.

(b) Extend the calculation of the solution for a time period $T = 60$ s after the end of the irradiation, by setting the rate of production of electron–hole pairs equal to $R = 0$. Graph the final concentration of electrons $n(t + T)$ at the end of the 60 s relaxation period as a function of irradiation time t. Show that this relaxation procedure results in a linear function $n(t + T)$ for the filling of the traps.

Solution

The following program in *Mathematica* solves the system of differential equations (4.4)–(4.7) and graphs the solution represented by the functions $n(t)$ and $n(t + T)$.

The main program **programMain** contains the two subroutines, **solveDiffeq** and **initValues,** which are used in several of the *Mathematica* programs in this chapter. The first subroutine **solveDiffeq** solves the system of four differential equations above with the initial conditions given, and stores the result of the numerical integration as the parameter **sol**. The command **NDSolve** is used once more to perform the numerical integration of the system of coupled differential equations. The subroutine **solveDiffeq** is called by using a total of six parameters, namely the initial condition parameters $n(0)$, $n_h(0)$, $n_v(0)$, and $n_c(0)$ (represented by the variables n10, nh0, nv0, nc0), the pair production parameter **R** and the irradiation time **tfinal**.

The second subroutine **initValues** sets the initial values of $n(0)$, $n_h(0)$, $n_v(0)$, and $n_c(0)$ at the beginning of the *relaxation stage* equal to the final values of $n(t)$, $n_h(t)$, $n_v(t)$, and $n_c(t)$ at the end of the *irradiation stage*. This subroutine is called by using two parameters, the parameter **b** (representing the solution of the system of differential equations) and the irradiation time represented by the parameter **d**.

There are two stages in the simulation, the *irradiation stage* and the *relaxation stage*. During the irradiation stage the subroutine **solveDiffeq** is called to solve the system of differential equations for a certain irradiation time given by the parameter **tfinal9=irrTime**, and for the given value of $R = 10^{14}\,\text{cm}^{-3}\,\text{s}^{-1}$. The solution is stored in the parameter **sol9**. At the end of the irradiation stage the subroutine **initValues** is called in order to set-up the initial values for the subsequent relaxation stage, as described above.

Next, during the relaxation stage, the subroutine **solveDiffeq** is called to solve the system of differential equations for a time period T=**tfinal10**=60 s and for

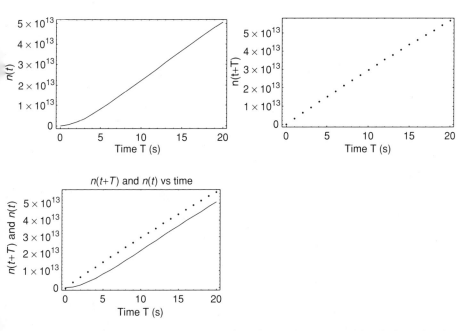

FIGURE 4.2. Results of the model for the filling of traps during crystal irradiation using Mathematica.

the value of $R = 0$. This 60-s relaxation period allows the electrons and holes accumulated in the conduction and valence band to be trapped into the available trap and recombination centers. The solution of the differential equations for this stage is stored in the parameter **sol10**.

The program contains a **FOR** loop, which calls the **programMain** and solves the system of differential equations for several irradiation times. The various irradiation times are contained in the parameter **irrTime**, starting at 0 s and ending at 20 s, in steps of 1 s. The values of $n(t)$ at the end of each irradiation stage are saved in the *Mathematica* list **n1List1** by using the command **AppendTo**. In a similar manner the values of $n(t + T)$ at the end of the irradiation stage are saved in the *Mathematica* list **n1List2**. Finally the two lists are graphed by using the command **ListPlot**.

The result of running the program is shown in Figure 4.2.

The first graph shows that a nonlinear function $n(t)$ for the filling of the traps is obtained at the end of the irradiation period. The second graph indicates that the final function $n(t + T)$ calculated at the end of the 60 s relaxation period, yields a linear function for the filling of the traps. The last graph is identical to the published data in Chen et al [10], Figure 2.

This exercise illustrates the importance of including appropriate relaxation periods after each irradiation or heating stage of multistage simulations of the TL kinetic processes.

One can also obtain valuable insight into the nature of the trap filling process in this example, by further examining the variation of $n(t)$, $n_h(t)$, $n_v(t)$, and $n_c(t)$

Irradiation graphs for irradiation time = 10 sec

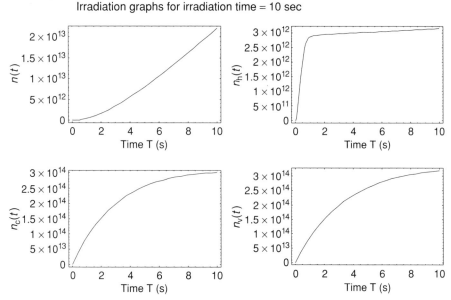

FIGURE 4.3. Variation of the functions $n(t)$, $n_h(t)$, $n_c(t)$, and $n_v(t)$ during the irradiation stage.

during the irradiation and relaxation process. This can be accomplished easily by using the **ListPlot** command during each stage of the simulation.

Typical results are shown in Figure 4.3 for the variation of the functions of $n(t)$, $n_h(t)$, $n_c(t)$, and $n_v(t)$ during the irradiation stage, for a total irradiation period of $t = 10$ s. These graphs show that the functions $n_h(t)$, $n_v(t)$, and $n_c(t)$ increase with time during the irradiation process, indicating that holes and electrons are accumulating in the valence and conduction band.

Figure 4.4 shows the variation of the same functions during the subsequent 60-s relaxation stage. It is seen that the concentrations of holes in the valence band (n_v) and electrons in the conduction band (n_c) decrease during the 60-s relaxation period, and that they quickly reach negligible values. The electrons and holes released from the bands get trapped in the electron trap and in the recombination center correspondingly, resulting in an increase of the final concentrations $n(t)$ and $n_h(t)$. At all times, the sum total of electron concentrations $n(t) + n_c(t)$ is equal to the sum total of the hole concentrations $n_v(t) + n_h(t)$, as required by charge conservation.

Listing of Program for Exercise 4.1

```
Remove["Global`*"];
programMain := (
  A = 10^-17; Ar = 10^-13; Ah = 10^-15;
  N1 = 10^15; Nh = 3*10^14;
```

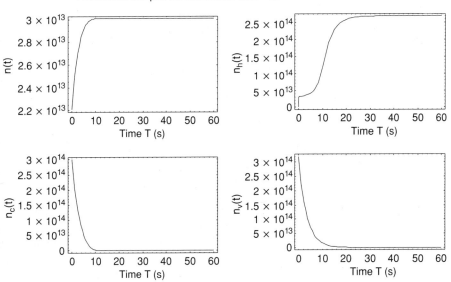

FIGURE 4.4. Variation of $n(t)$, $n_h(t)$, $n_c(t)$, and $n_v(t)$ during the 60-s relaxation stage.

```
solveDiffeq[n10_,nh0_,nv0_,nc0_,R_,tfinal_]:=Module[{t},
    sol = NDSolve[{nl'[t] == A*(N1-nl[t])*nc[t],
        nh'[t] == -Ar*nh[t]*nc[t] + Ah*nv[t]*(Nh-nh[t]),
        nv'[t] == R-Ah*nv[t]*(Nh-nh[t]), nc'[t] == nh'[t]
        + nv'[t] -nl'[t], nl[0] == n10,
        nc[0] == nc0, nv[0] == nv0, nh[0] == nh0, {nl, nh, nv,
        nc, {t, 0, tfinal}, MaxSteps → 50000]];
initValues[b_,d_]: = Module[{},n10 = Last[nl[d]/.b];
    nh0 = Last [nh[d]/.b]; nv0 = Last[nv[d]/.b];
    nc0 = Last[nc[d]/.b]];

(*-------------------------------------------------------*)

(*irradiation *)
R = 10^14; tfinal9 = irrTime;
solveDiffeq[n10, nh0, nv0, nc0, R, tfinal9];
Sol9=sol;

(* Relaxation stage *)
initValues[sol9,tfinal9];
R = 0; tfinal10 = 60;
solveDiffeq[n10, nh0, nv0, nc0, R, tfinal10];
sol10 = sol;
)
(*-----------------------------------------------------*)

tstart = 0; tend = 20; tstep = 1;
```

```
nlList1 = {}; nlList2 = {};
For[tloop = tstart, tloop ≤ tend, tloop+= tstep,
  n10 = 0; nh0 = 0; nv0 = 0; nc0 = 0;
  irrTime = tloop;
  programMain;
  initValues[sol9, tfinal9]; (*find n(t) at the end of
    irradiation stage*);
  AppendTo[nlList1, {tloop, n10}];
  initValues[sol10, tfinal10]; (*find n(t+T) at the end of
    relaxation stage*);
  AppendTo[nlList2, {tloop, n10}];
]

gr1 = ListPlot[nlList1, PlotRange→All, PlotJoined → True,
  PlotLabel→ "n(t) vs t", ImageSize → 723];
gr2 = ListPlot[nlList2, PlotRange→All, PlotJoined→True,
  PlotLabel→ "n(t+T) vs t", ImageSize → 723];
Show[{gr1, gr2}, PlotLabel→ "n(t+T) and n(t) vs time"];
```

Exercise 4.2: Competition During Excitation Model

Write a computer program to integrate the kinetic rate equations relevant to the TL model shown in Figure 4.5 [11]. The model consists of two electron trapping states characterized by total concentrations N_1 and N_2, and by instantaneous occupancies $n_1(t)$ and $n_2(t)$, respectively. The first trap is considered to be the one responsible for TL, and the second trap is denoted as the competitor trap. The model also contains a recombination center, with instantaneous hole occupancy $p(t)$.

During the irradiation process the electrons are raised from the valence band into the conduction band, and can be trapped into either N_1 or N_2, with the two traps competing for the electrons as shown in Figure 4.5.

The kinetic equations for this model are [11]

$$\frac{dn_1}{dt} = A_1(N_1 - n_1)n_c \tag{4.8}$$

$$\frac{dn_2}{dt} = A_2(N_2 - n_2)n_c \tag{4.9}$$

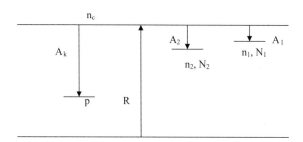

FIGURE 4.5. The competition during excitation model.

$$\frac{dn_c}{dt} = R - \frac{dn_1}{dt} - \frac{dn_2}{dt} - A_k n_c p \qquad (4.10)$$

$$p = n_1 + n_2 + n_c. \qquad (4.11)$$

The first two equations express mathematically the fact that electrons in the conduction band can be trapped into either the main or into the competitor trap. The third equation expresses the fact that the electrons in the conduction band are produced by the constant excitation rate R, and can also be trapped into either of the two traps (terms $-\frac{dn_1}{dt}$ and $-\frac{dn_2}{dt}$), or into the recombination center (term $-A_k n_c p$). The last equation (4.11) expresses the conservation of total charge in the crystal, with the left-hand side being equal to the concentration of holes trapped in the recombination center, and the right-hand side representing the total concentration of electrons in the crystal at any moment t. As discussed in the book by Chen and McKeever [1], the last equation is based on the assumption that the concentration of free holes in the valence band can be neglected as compared with the accumulated concentration of holes $p(t)$.

The parameters in the above expressions are as follows:

$A_1 =$ transition probability coefficient of electrons into the main trap $(m^3\,s^{-1})$

$A_2 =$ transition probability coefficient of electrons into the competitor trap $(m^3\,s^{-1})$

$A_k =$ transition probability coefficient of electrons from the conduction band into the recombination center $(m^3\,s^{-1})$

$n_1 =$ instantaneous concentration of electrons in the main trap at time $t\,(m^{-3})$

$N_1 =$ total concentration of main traps in the crystal (m^{-3})

$(N_1 - n_1) =$ instantaneous concentration of empty main traps available at time t

$n_2 =$ instantaneous concentration of electrons in the competitor trap (m^{-3})

$N_2 =$ total concentration of competitor traps in the crystal (m^{-3})

$n_c =$ instantaneous concentration of electrons in the conduction band (m^{-3})

$R =$ constant rate of production of electron–hole pairs per m^3 per second $(m^{-3}\,s^{-1})$

$p =$ instantaneous concentration of holes in the recombination center (m^{-3})

Use the following numerical values $N_1 = N_2 = 10^{23}\,m^{-3}$, $A_1 = A_k = 10^{-22}\,m^3\,s^{-1}$, and $A_2 = 3 \times 10^{-21}\,m^3\,s^{-1}$, $R = 10^{21}\,m^{-3}\,s^{-1}$. The initial conditions at time $t = 0$ are $n_1(0) = n_2(0) = n_c(0) = p(0) = 0$.

Solution

The following program in *Mathematica* solves the system of differential equations (4.8)–(4.10) and graphs the solution represented by the functions $n_1(t)$, $n_2(t)$, and

$n_c(t)$. The structure of the program is very similar to that of the previous exercise, with the main program **programMain** containing the two subroutines **solveDiffeq** and **initValues**. The first subroutine **solveDiffeq** is called by using a total of five parameters: the three initial condition parameters $n_1(0)$, $n_2(0)$, and $n_c(0)$, the pair production parameter **R**, and the irradiation time **tfinal**. The second subroutine **initValues** sets the initial values of $n_1(0)$, $n_2(0)$, and $n_c(0)$ at the beginning of the relaxation stage equal to the final values of $n_1(t)$, $n_2(t)$, and $n_c(t)$ at the end of the irradiation stage.

The program contains a **FOR** loop, which solves the system of differential equations for several irradiation times contained in the parameter **irrTime**, starting at 1 s and ending at 800 s, in steps of 30 s. The values of $n_1(t)$, $n_2(t)$, $n_c(t)$, and $\ln(n_1(t))$ at the end of the irradiation stage are saved in the *Mathematica* lists **n1List1**, **n2List1**, **ncList**, and **logn1List1** and are graphed by using the command **ListPlot**.

The result of running the program is shown in Figure 4.6.

The graphs of $n_1(t)$ above shows the existence of two regions with different dose response, for $0 < t < 150$ s and for $t > 150$ s. The change in TL dose response coincides with the onset of saturation effects for the competitor trap at approximately $t = 150$ s, as seen clearly in the second graph.

The regions of different dose response are seen clearly in the log–log plot of $n_1(t)$ versus time, which shows the existence of an initial linear dose response on the log–log scale, followed by a superlinear region and finally the onset of saturation for the main trap.

The last graph is identical to the published data in Bowman and Chen [11], Figure 2.

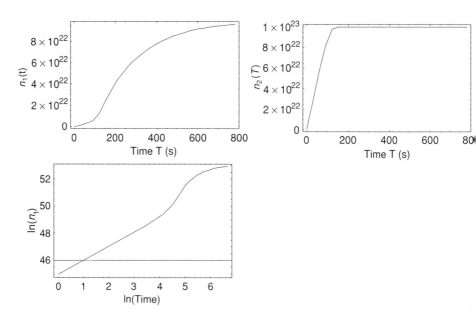

FIGURE 4.6. Results from the competition during the excitation model.

Listing of Program for Exercise 4.2

```
Remove["Global`*"];
programMain := (
  k1 = 8.617*10^-5;
  A1 = 10^-22; A2=30*10^-22; Ak = 10^-22;
  N1 = 10^23; N2 = 10^23;

  solveDiffeq[n10_,n20_,nc0_,R_,tfinal_1]:=
  Module[{t}, sol =
    NDSolve[{n1'[t] == A1*(N1-n1[t])*nc[t],
      n2'[t] == A2*(N2-n2[t]*nc[t], nc'[t] == R-n1'[t]-n2'
        [t]-Ak*nc[t]*(n1[t] + n2[t] + nc[t]),n1[0] == n10,
      n2[0] == n20,nc[0] == nc0}, {n1,n2,nc}, {t,0,tfinal},
      MaxSteps→50000]];

  initValues[b_,d_] : = Module[{},n10 = Last[n1[d]/.b];
    n20 = last[n2[d]/.b];
    nc0 = Last[nc[d]/.b]];

  (*-------------------------------------------------*)

  (*irradiate for irrTime*)
  R = 10^21; tfinal7 = irrTime;
  solveDiffeq[n10,n20,nc0,R,tfinal7];
  sol7 = sol;
  graphAllEleven[sol7, tfinal7];

  (*Relaxation stage *)
  initValues[sol7,tfinal7];
  R = 0; tfinal8 = 60;
  SolveDiffeq[n10,n20,nc0,R, tfinal8];
  sol8 = sol;
  graphAllEleven[sol8, tfinal8];
  )
(*-------------------------------------------------*)

tstart = 1; tend = 800;tstep = 30;
n1List1 = {}; n2List1 = {};ncList1 = {}; logn1List1 = {};
For[tloop = tstart, tloop ≤ tend, tloop,+ = tstep,
  irrTime=tloop;n10=0;n20=0;nc0=0;
  programMain;
  initValues[sol8, tfinal8]; (*find n1,n2,nc at end of
    relaxation section*);
  AppendTo[n1List1, {irrTime,n10}]; AppendTo [n2List1,
    {irrTime, n20}];
  AppendTo[ncList1, {irrTime,nc0}]; AppendTo [logn1List1,
    {Log[irrTime], Log[n10]}];
]
ListPlot[n1List1, PlotRange → All, PlotJoined → True,
```

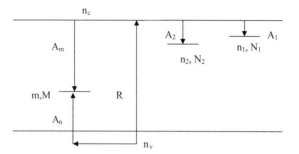

FIGURE 4.7. Model with competition during both excitation and heating: irradiation stage.

```
PlotLabel  →  "n1 vs t", ImageSize  →  723];
ListPlot[n2List1, PlotRange  →  All, PlotJoined  →  True,
    PlotLabel  →  "n2 vs t", ImageSize  →  723];
ListPlot[logn1List1, PlotRange  →  All, PlotJoined  →  True,
    PlotLabel  →  "ln(n1) vs ln (t)",ImageSize  →  723];
```

Exercise 4.3: Superlinearity Model with Competition During Both Excitation and Heating

Write a computer program to integrate the kinetic rate equations relevant to the TL model shown in Figures 4.7 and 4.8 [9]. The model consists of two trapping states characterized by concentrations N_1 and N_2, and by instantaneous occupancies $n_1(t)$ and $n_2(t)$ respectively. The first trap is considered to be the one responsible for TL, and the second trap is denoted as the competitor trap. The model also contains a recombination center, with instantaneous occupancy $m(t)$ and total concentration of hole traps given by M.

The simulation should contain three stages: the irradiation process, an intermediate relaxation stage, and the heating (measurement of TL) stage.

During the irradiation process the electrons are raised from the valence band into the conduction band, and can be trapped into either N_1 or N_2, with the two traps competing for the electrons. These electrons in the conduction band can

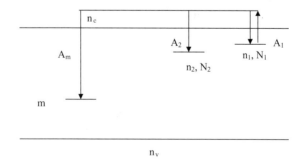

FIGURE 4.8. Model with competition during both excitation and heating: heating stage.

also recombine with holes at the recombination center. Simultaneously, an equal number of holes are created in the valence band by the irradiation process. These holes can be trapped directly into the recombination center increasing the hole occupancy $m(t)$.

The kinetic equations for the excitation stage in this model are [9]

$$\frac{dn_1}{dt} = A_1(N_1 - n_1)n_c \tag{4.12}$$

$$\frac{dn_2}{dt} = A_2(N_2 - n_2)n_c \tag{4.13}$$

$$\frac{dm}{dt} = A_n n_v(M - m) - A_m m n_c \tag{4.14}$$

$$\frac{dn_v}{dt} = R - A_n n_v(M - m) \tag{4.15}$$

$$\frac{dn_c}{dt} + \frac{dn_1}{dt} + \frac{dn_2}{dt} = \frac{dm}{dt} + \frac{dn_v}{dt}. \tag{4.16}$$

The first two equations express mathematically the fact that electrons in the conduction band can be trapped into either the main or the competitor trap. The third equation expresses the fact that the number of holes in the recombination center is changed by either trapping additional holes from the valence band (term $A_n n_v(M - m)$), or by trapping electrons from the conduction band (term $-A_m m n_c$). The fourth equation expresses the fact that holes are produced continuously in the valence band by the excitation rate R, but they are also captured into the recombination center (term $-A_n n_v(M - m)$). The last equation is the conservation of total charge in the crystal, with the left-hand side being equal to the total rate of change of the concentration of holes, and the right-hand side being equal to the total rate of change of the concentration of electrons in the crystal.

The parameters in the above expressions are as follows:

$A_1 =$ transition probability coefficient of electrons into the main trap $(m^3 \, s^{-1})$

$A_2 =$ transition probability coefficient of electrons into the competitor trap $(m^3 \, s^{-1})$

$A_m =$ transition probability coefficient of electrons from the conduction band into the recombination center $(m^3 \, s^{-1})$

$A_n =$ capture probability coefficient of holes from the valence band into the recombination center $(m^3 \, s^{-1})$

$n_1 =$ instantaneous concentration of electrons in the main trap at time t (m^{-3})

$N_1 =$ total concentration of main traps in the crystal (m^{-3})

$(N_1 - n_1) =$ instantaneous concentration of empty main traps available at time t

$n_2 =$ instantaneous concentration of electrons in the competitor trap (m^{-3})

$N_2 =$ total concentration of competitor traps in the crystal (m^{-3})

n_c = instantaneous concentration of electrons in the conduction band (m^{-3})

n_v = instantaneous concentration of holes in the valence band (m^{-3})

R = constant rate of production of electron–hole pairs per m^3 per second $(m^{-3}\,s^{-1})$

m = instantaneous concentration of holes in the recombination center (m^{-3})

M = total concentration of holes in the crystal (m^{-3})

Use the following numerical values $N_1 = 10^{23}\,m^{-3}$, $N_2 = 10^{21}\,m^{-3}$, $M = 1.01 \times 10^{23}\,m^{-3}$, $A_1 = 10^{-21}\,m^3 s^{-1}$, $A_2 = 10^{-19}\,m^3\,s^{-1}$, $A_m = 10^{-21}\,m^3\,s^{-1}$, $A_n = 10^{-21}\,m^3\,s^{-1}$, and $R = 10^{21}\,m^{-3}\,s^{-1}$. The initial conditions at time $t = 0$ are $n_1(0) = n_2(0) = n_c(0) = n_v(0) = m(0) = 0$.

The irradiation stage must be followed by a relaxation period of 60 s, to allow the charges in the conduction and valence band to relax into the various energy levels. During the heating stage the electrons are raised thermally from the main trap into the conduction band with a probability equal to $s\,\exp(-E/kT)$, and they can be trapped into either N_1 or N_2, with the two traps competing for the electrons. The competing trap is assumed to be thermally disconnected, so that there is no thermal excitation of the trapped electrons (n_2) into the conduction band. The electrons can also recombine with holes in the recombination center to produce the observed TL. After the end of the irradiation no additional holes are being produced in the valence band, so we can assume that $n_v = 0$ at all times during the heating stage.

Figure 4.8 shows the possible transitions during the heating stage.

The kinetic equations for the heating stage in this model are [9]

$$\frac{dn_1}{dt} = -sn_1\exp(-E/kT) + A_1(N_1 - n_1)n_c \qquad (4.17)$$

$$\frac{dn_2}{dt} = A_2(N_2 - n_2)n_c \qquad (4.18)$$

$$\frac{dn_v}{dt} = n_v = 0 \qquad (4.19)$$

$$I = -\frac{dm}{dt} = A_m m n_c \qquad (4.20)$$

$$\frac{dm}{dt} = \frac{dn_c}{dt} + \frac{dn_1}{dt} + \frac{dn_2}{dt} \qquad (4.21)$$

The additional parameters in these equations are E = activation energy and s = frequency factor for the main trap.

Solution

The following program in *Mathematica* solves the system of differential equations (4.12)–(4.21) and graphs the solution represented by the functions $n_1(t)$, $n_2(t)$, $m(t)$, $n_v(t)$, and $n_c(t)$. The structure of the program is very similar to that of the previous exercise, with the main program **programMain** containing the two

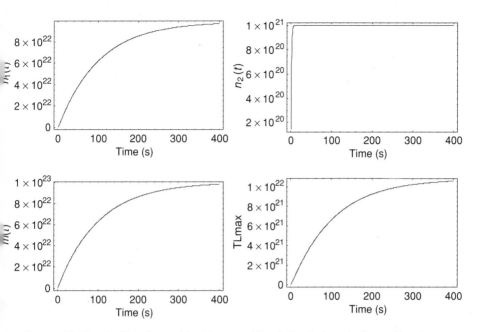

FIGURE 4.9. Results from the model with competition during both excitation and heating.

subroutines, **solveDiffeq** and **initValues**. The first subroutine **solveDiffeq** is called by using a total of nine parameters: the five initial condition parameters $n_1(0), n_2(0)$, $n_v(t), m(0)$, and $n_c(0)$, the pair production parameter **R**, the relevant time **tfinal**, the irradiation temperature **irrTemp**, and the heating rate parameter β**heat**. The second subroutine **initValues** provides the connection between different stages by setting the initial values of $n_1(0), n_2(0), m(0), n_v(t)$, and $n_c(0)$ at the beginning of each stage equal to the final values of $n_1(t), n_2(t), m(t), n_v(t)$ and $n_c(t)$ at the end of the previous stage.

The subroutines are called three times, for the irradiation, relaxation, and heating stages. It is noted that if irradiation of the sample is taking place at room temperature, the thermal excitation probability $s \exp(-E/kT)$ is equal to zero. As a result of this, we can combine the two sets of differential equations above, as seen in the listing of the Mathematica program. The program also calculates and saves the maximum TL intensity (TL_{max}) in the parameter sn.

The program contains a **FOR** loop, which solves the system of differential equations for several irradiation times contained in the parameter **irrTime**, starting at 0.3 s and ending at 400 s, in steps of 0.5 s. The values of $n_1(t), n_2(t), n_c(t), m(t)$, $n_v(t)$, and $\ln(n_1(t))$ at the end of the relaxation stage, as well as their logarithms, are saved in *Mathematica* lists and are graphed by using the command **ListPlot**.

The result of running the program is shown in Figure 4.9.

Figure 4.10 shows the functions $n_1(t), n_2(t), m(t)$, and TL_{max} as a function of the irradiation time t, on a log–log scale. It is seen that at small irradiation times the TL versus dose graph is initially quadratic (slope = 2 on the log–log scale),

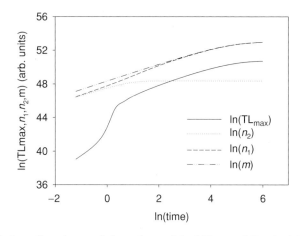

FIGURE 4.10. Superlinearity graph from the model of Chen and Fogel with competition taking place during both the excitation and heating stages.

followed by a superlinear region where $k > 2$, to be followed by a linearity region (slope $k = 1$), and finally by approach to saturation.

Figure 4.10 is identical to the published data of Chen and Fogel [9].

Listing of Program for Exercise 4.3

```
Remove["Global`*"];
programMain : = (
  E1 = 1; s1 = 10^13;
  k1 = 8.617*10^-5;
  f = 10^21;
  A1 = 10^-21;A2 = 10^19; Am = 10^-21;Ah = 10^-21;
  N1 = 10^23;N2 = 10^21;M = 1.01*10^23;

  solveDiffeq[n10_, n20_, m0_, nv0_, nc0_, R_, tfinal_,
      βheat_, irrTemp_] : =
    Module[{t},
     sol =
      NDSolve[{n1'[t] == A1*(N1-n1[t])*nc[t]-
        n1[t]*s1*E^ (-E1/(k1*(273+irrTemp+βheat*t))),
        n2'[t] == A2*(N2-n2[t])*nc[t],
        m'[t] == -Am*m[t]*nc[t]+Ah*nv[t]*([M-m[t]),
        nv'[t] == R-Ah*nv[t]*(M-m[t]), nc'[t] == m'[t]
        + nv'[t]-n1' [t]-n2'[t],n1[0] == n10,
        n2[0] == n20,nc[0] == nc0, nv[0] == nv0,m[0]
        == m0},{n1,n2,m,nv,nc}, {t,0,tfinal},
      MaxSteps→50000]];

initValues[b_,d_]:= Module[{},n10 = Last[n1[d]/.b];
```

```
  n20 = Last[n2[d]/.b];m0 = Last[m[d]/.b];
  nv0 = Last[nv[d] /.b];nc0 = Last[nc[d] /.b]];

(*---------------------------------------------*)

(* irradiation *)
R = 10^21;tfinal19 = irrTime; βheat = 0; irrTemp = 20;
solveDiffeq[n10, n20,m0,nv0,nc0,R,tfinal9,βheat,
irrTemp]; sol9=sol;

(* Relaxation stage *)
initValues[sol9,tfinal9];
R = 0;tfinal10 = 60; βheat = 0; irrTemp = 20;
solveDiffeq[n10, n20,m0,nv0,nc0,R,tfinal10,βheat,
irrTemp];sol10=sol;

(*step 3- record maxTL*)
initValues[sol10,tfinal10];
R = 0;βheat = 5; irrTemp = 20;tfinal11 = (170-irrTemp)/
  βheat;
solveDiffeq[n10, n20,m0,nv0,nc0,R,tfinal11,βheat,
  irrTemp]; sol11 = sol;
tlList4 = Table[{First[Evaluate[(m[t]*nc[t]*Am)/
        .sol11]]},{t,0,tfinal11,0.2}];
sn = Max[tlList4];
)
(*---------------------------------------------*)
tstart = 0.3;tend = 400;tstep = 0.5;
n1List = {}; n2List1 = {}; mList1 = {}; snList1 = {};
 logsnList1 = {}; logn1List1 = {}; logn2List1 = {};
 logmList1 = {};
For[tloop=tstart, tloop ≤ tend,tloop += tstep,
 n10=0;n20=0;m0=0;nv0=0;nc0=0;
 irrTime=tloop;
 programMain;
 initValues[sol10,tfinal10]; (*find n1,n2,m,TLmax at
   end of first irradiation section*);
 AppendTo[lognList1,{Log[irrTime], Log[n10]}];
   AppendTo [logn2List1,{Log[irrTime], Log[n20]}];
 AppendTo[logmList1,{Log[irrTime], Log[m0]}]; AppendTo
   [logsnList1,{Log[irrTime], Log[sn]}];
 AppendTo[n1List1,{tloop, n10}]; AppendTo[n2List1,
   {tloop, n20}]; AppendTo[mList1, {tloop, m0}];
 AppendTo[snList1, {tloop, sn}];
]

ListPlot[n1List1,PlotRange → All,PlotJoined → True,
 AxesLabel → {t,n1}, ImageSize → 723];
ListPlot[n2List1,PlotRange → All,PlotJoined → True,
```

TABLE 4.1. The TL response vs Dose D data

Dose (Gy)	$y(D)$ (a.u.)	Dose (Gy)	$y(D)$ (a.u.)
0.001	0.001166	75	170.2
0.005	0.005831	100	256.5
0.010	0.01166	250	942.1
0.050	0.05834	500	2,204
0.100	0.1168	750	3,204
0.500	0.5873	1,000	3,876
1	1.183	2,000	4,753
2	2.4	5,000	4,844
5	6.254	7,500	4,844
10	13.34	10,000	4,844
25	39.32	50,000	4,844
50	96.97		

```
AxesLabel → {t,n2}, ImageSize → 723];
ListPlot[mList1,PlotRange → All,PlotJoined → True,
AxesLabel → {t,m}, ImageSize → 723];
ListPlot[snList1,PlotRange → All,PlotJoined → True,
AxesLabel → {t,TLmax}, ImageSize → 723];
ListPlot[logsnList1,PlotRange → All,PlotJoined → True,
AxesLabel → {ln(t),ln(TLmax)}, ImageSize → 723];
ListPlot[logn1List1,PlotRange → All,PlotJoined → True,
AxesLabel → {ln(t),ln(n1)}, ImageSize → 723];
ListPlot[logn2List1,PlotRange → All,PlotJoined → True,
AxesLabel → {ln(t),ln(n2)}, ImageSize → 723];
ListPlot[logmList1,PlotRange → All,PlotJoined → True,
AxesLabel → {ln(t),ln(m)}, ImageSize → 723];
```

Exercise 4.4: The $f(D)$ and $g(D)$ Functions

Table 4.1 lists the gamma doses (D) given to a set of thermoluminescence dosimeters (TLDs), and the corresponding TL net emission, $y(D)$ in arbitrary TL reader units.

Figure 4.11 shows the plot $y(D)$ versus D obtained from the data. The experimental data have been fitted by the following equation

$$y(D) = A(1 - e^{-BD}) - CDe^{-BD} \qquad (4.22)$$

where D is the given dose, A is the TL response at the saturation level (4844 a.u.), $B = 0.291 \times 10^{-2}$ Gy^{-1}, and C = 12.93 Gy^{-1}.

The aim of this exercise is to study the functions $f(D)$ and $g(D)$ and determine for which doses the TL response is linear, superlinear, supralinear, or sublinear.

The superlinearity index $g(D)$ gives an indication of the change in the slope of the dose response; the supralinearity index $f(D)$ is used in the case where the main interest is in the amount of deviation from linearity, i.e., whether the TL signal is above or below the linear extrapolated range, and to make corrections if needed.

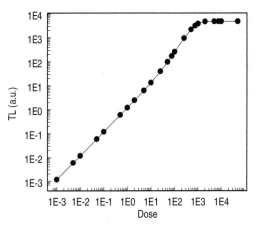

FIGURE 4.11. Plot of the TL response versus dose.

The superlinearity index $g(D)$ is defined as

$$g(D) = \left[\frac{D y''(D)}{y'(D)} \right] + 1 \qquad (4.23)$$

while the supralinearity index $f(D)$ is defined as

$$f(D) = \frac{\left[\dfrac{y(D)}{D} \right]}{\left[\dfrac{y(D_1)}{D_1} \right]}. \qquad (4.24)$$

The following is the meaning of the symbols used in the two previous equations:

- y denotes the TL signal obtained at a given dose D;
- $y = y(D)$ is the analytical expression giving the behavior of y as a function of dose D and it is found by fitting the experimental values;
- $y'(D)$ and $y''(D)$ are, respectively, the first and second derivatives of the function $y(D)$;
- D_1 is the normalization dose in the initial linear range of the $y = y(D)$ curve.

Solution

The first and the second derivatives of equation (4.22) are

$$y' = (A \cdot B - C + B \cdot C \cdot D) e^{-BD} \qquad (4.25)$$

$$y'' = (2C - A \cdot B - B \cdot C \cdot D) B e^{-BD}. \qquad (4.26)$$

Table 4.2 gives the numerical values of the first and second derivative as a function of the dose D.

Calculating now the $g(D)$ and $f(D)$ functions according to equations (4.23) and (4.24) and plotting both the first and second derivative as a function of the dose, we obtain Figures 4.12 and 4.13, respectively.

TABLE 4.2. The values of the first and second derivatives as a function of dose

Dose (Gy)	$y'(D)$	$y''(D)$	Dose (Gy)	$y'(D)$	$y''(D)$
0.001	1.166	0.03423	75	3.206	0.02092
0.005	1.166	0.03423	100	3.684	0.01741
0.010	1.166	0.03423	250	5.108	0.003314
0.050	1.168	0.03422	500	4.663	-0.4788×10^{-2}
0.100	1.169	0.03421	750	3.314	-0.5400×10^{-2}
0.500	1.183	0.03413	1,000	2.113	-0.4100×10^{-2}
1	1.2	0.03402	2,000	0.2268	-0.5483×10^{-3}
2	1.234	0.03382	5,000	9.08×10^{-3}	-0.2462×10^{-6}
5	1.335	0.03320	7,500	9.41×10^{-7}	-0.2615×10^{-9}
10	1.498	0.03219	10,000	8.68×10^{-11}	-3.2824×10^{-13}
25	1.959	0.02929	50,000	1.215×10^{-60}	-3.7499×10^{-63}
50	2.635	0.02486			

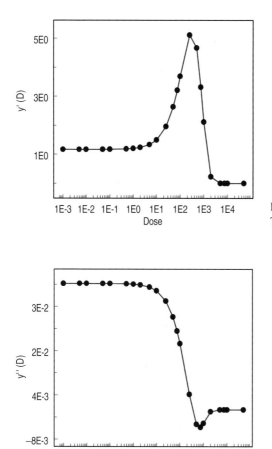

FIGURE 4.12. First derivative of the TL signal $y(D)$.

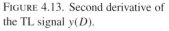

FIGURE 4.13. Second derivative of the TL signal $y(D)$.

TABLE 4.3. The $f(D)$ and $g(D)$ functions Vs dose

Dose (Gy)	$g(D)$	$f(D)$	Dose (Gy)	$g(D)$	$f(D)$
0.001	1	1	75	1.489	1.946
0.005	1	1	100	1.472	2.200
0.010	1	1	250	1.182	3.232
0.050	1.001	1.001	500	0.4866	3.781
0.100	1.003	1.001	750	-0.2222	3.664
0.500	1.014	1.007	1,000	-0.9401	3.324
1	1.028	1.015	2,000	-0.3835×10^1	2.038
2	1.055	1.029	5,000	-0.1256×10^2	0.8308
5	1.124	1.073	7,500	-0.1983×10^2	0.5539
10	1.215	1.144	10,000	-0.3680×10^2	0.4154
25	1.347	1.349	50,000	-0.1532×10^3	0.08307
50	1.472	1.663			

Table 4.3 lists the $f(D)$ and $g(D)$ functions versus dose.

We now plot both functions $f(D)$ and $g(D)$ as a function of the dose, as shown in Figures 4.14 and 4.15.

In Figure 4.16 we show both coefficients $f(D)$ and $g(D)$ on the same graph, to emphasize their differences and similarities. It is noted that $g(D)$ is a functional representation of the nonlinear response and therefore can have negative values as well.

On the other hand, $f(D)$ is a normalization of the data with respect to the lowest available dose and therefore can only have positive values.

Discussion

Values of $g(D) > 1$ indicates a superlinearity region, while $g(D) = 1$ means a linear region and $g(D) < 1$ denotes a sublinearity region.

In a similar manner, $f(D) < 1$ means a sublinear region, $f(D) \sim 0$ means a saturation region and $f(D) > 1$ indicates a supralinearity region.

The general features concerning the TL versus dose behavior can be outlined as follows:

- if $y''(D) > 0$, $y'(D)$ and $y(D)$ increase with D and $y(D)$ is superlinear,
- if $y''(D) < 0$, $y'(D)$ and $y(D)$ decrease with D and $y(D)$ is sublinear,
- if $y''(D) = 0$, $y'(D)$ is constant with D and $y(D)$ is linear.

The analysis of some of the data in Figure 4.11 indicates the following results.

D = 50 Gy: $y' > 0$ means y is increasing; $y'' > 0$ and y' is increasing, and the graph is concave upwards. $g > 1$ and so y is superlinear; $f > 1$ means the curve is also supralinear.

D = 500 Gy: $y' > 0$, then y is increasing; $y'' < 0$, y' is decreasing and the graph is concave downwards; $g < 1$ means y is sublinear; f > 1 means y is supralinear.

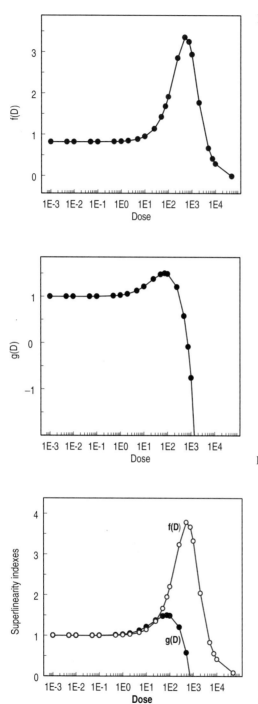

FIGURE 4.14. The function $f(D)$.

FIGURE 4.15. The function $g(D)$.

FIGURE 4.16. The functions $g(D)$ and $f(D)$ are graphed together.

$\mathbf{D} = \mathbf{10^4 Gy}$: $y' > 0$ which means y is increasing; because $y'' < 0$, y' is decreasing and the graph is concave downwards; $g < 1$, $f < 1$ mean y is sublinear in D, and y is approaching saturation.

References

[1] R. Chen and S.W.S. McKeever 1997. *Theory of Thermoluminescence and Related Phenomena.* Singapore: World Scientific, Chapter 4.

[2] S.W.S. McKeever and R. Chen, *Radiat. Meas.* **27**, (1997) 625.

[3] A. Halperin and R. Chen, *Phys. Rev.* **148** (1966) 839.

[4] R. Chen and S.W.S. McKeever, *Radiat. Meas.* **23** (1994) 667.

[5] J.R. Cameron, N. Suntharalingam, and G.N. Kenney 1968. *Thermoluminescent Dosimetry.* Madison: The University of Wisconsin Press.

[6] E.F. Mische and S.W.S. McKeever, *Radiat. Prot. Dosim.* **29** (1989) 159.

[7] E.T. Rodine and P.L. Land, *Phys. Rev.* **B 4** (1971) 2701.

[8] N. Kristianpoller, R. Chen, and M. Israeli. *J. Phys. D: Appl. Phys.* **7** (1974) 1063.

[9] R. Chen and G. Fogel. *Radiat. Prot. Dosim.* **47** (1993) 23.

[10] R. Chen, S.W.S. McKeever, and S.A. Duranni, *Phys. Rev.* B **24** (1981) 4931.

[11] S.G.E. Bowman and R. Chen, 1979, *J. Lumin.* **18/19** (1970) 345.

5
Miscellaneous Applications
of Thermoluminescence

Introduction

The introductory sections of this chapter present some of the fundamental definitions, terminology, and equations used to describe the statistical accuracy and reproducibility of thermoluminescence data.

In Exercises 5.1–5.4, we study the accuracy and reproducibility of thermoluminescence dosimetry (TLD) systems and the various quantities used to describe them. Exercises 5.5 and 5.6 demonstrate two characteristic examples of data analysis in environmental dosimetry and dose monitoring.

In Exercises 5.7 and 5.8, we perform a numerical simulation of the phenomenon of thermal quenching that is exhibited by many TL materials.

In Exercise 5.9, the mathematical basis and a numerical simulation of the TL-like presentation of phosphorescence decay data are given. In Exercise 5.10, we look at the important experimental problem of temperature lag and how to correct experimental glow curves for this common experimental effect.

Finally, Exercise 5.11 is the study of the various terms appearing in the first-order and general-order equations for TL glow peaks. By performing a numerical analysis of these terms for a wide range of the activation energies E and the frequency factors s, a formula is derived for the activation energy E as a function of the full width at half maximum (FWHM) ω and the temperature T_M of maximum TL intensity. This formula is compared with the well-known equation (1.49).

Reproducibility of TLD Systems

The dosimetric properties of TLDs systems depend in a complex manner on the combination of several different factors, such as the individual TL equipment used in the experiment (usually referred to as the *TL reader*), the *TL dosimeters* used, and the *readout* and *annealing procedure* used for the dosimeters.

A quantity of interest in experimental TLD work is the *lower detection limit* D_{LDL} that is defined as three times the standard deviation σ_{BKG} of the zero-dose reading, which is a TL reading taken after annealing of the dosimeters but before any irradiation takes place. The lower detection limit D_{LDL} is typically given in units of the absorbed dose (Gy).

$$D_{LDL} = 3\sigma_{BKD}. \tag{5.1}$$

Burkhardt and Piesh [1] used a batch of 10 TLD dosimeters in a classic experimental study of the reproducibility of TLD systems. They irradiated these 10 dosimeters from the lowest detectable dose D_{LDL} of the system, up to a high dose equal to 1,000 times the D_{LDL}. They calculated the relative standard deviation $s(D)$ of the TL readings as a function of the dose D received by the 10 dosimeters, and presented it graphically as a function $s(D)$ of the dose D.

These authors found that the results could be represented by a two-parameter fit of the form:

$$s(D) = \left[\frac{A^2}{D^2} + B^2 \right]^{1/2}. \tag{5.2}$$

In this equation, the symbols represent the following:

$s(D)$ = relative standard deviation at different doses D
A = value of the absolute standard deviation (SD) at very low doses and
B = relative SD at high doses

A typical experimental result for the graph $s(D)$ as a function of dose D is shown in Figure 5.1, and represents graphically the reproducibility of dose measurements in TLD systems. The data in Figure 5.1 show that the relative standard deviation becomes very large at low doses, while it becomes essentially constant and of the order of a few percent at high doses.

The dose range where a constant $s(D)$ value is obtained depends on the properties of the batch of dosimeters used, as well as on the properties of the individual TL reader used in the experiment.

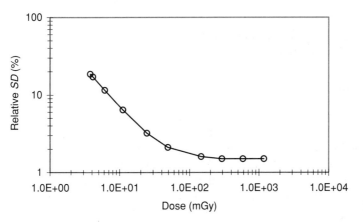

FIGURE 5.1. Typical reproducibility graph for TLD systems: the relative standard deviation $s(D)$ of the data obtained with a group of 10 dosimeters is plotted as a function of the dose D.

Zarand and Polgar ([2],[3]) developed a slightly different theoretical expression for the dependence of the relative standard deviation $s(D)$ on the dose D received by the dosimeter. They found that in some experimental situations the relative standard deviation of the data could be represented more accurately by the expression:

$$s(D) = \left[\frac{A^2}{D^2} + \frac{1}{kD} + B^2 \right]^{1/2}. \tag{5.3}$$

In this equation the extra term $1/kD$ appears, where D is the absorbed dose as in the previous equation and k is the constant representing the photoelectron-to-dose conversion factor. The rest of the symbols in this equation remain the same as in equation (5.2).

Another way of representing the reproducibility of a TLD system is by performing a statistical analysis using a group of dosimeters of the same type. For example, the group of 10 dosimeters are annealed, irradiated, and read out using the same procedure several times, perhaps 10 times for each dosimeter. The analysis of the means and standard deviations (also called the coefficients of variation) can help in identifying the different sources of variation in the reproducibility of the system. These sources can be associated with the TL reader, with the TL elements, or with some other source in the dosimetry procedure.

The following coefficients of variation can be defined [4] as:

The system variability index (SVI) or % $\overline{\text{CV}}$: this is the mean value of the percent standard deviations of each TL detector. This quantity gives a measure of the reproducibility of the whole system.

The reader variability index (RVI) or % $\overline{\text{CV}}$: this is the percent standard deviations of the mean values of each cycle of readings. This quantity gives a measure of the long-term reader reproducibility.

The detector variability index (DVI): this quantity gives a measure of the reproducibility of the TL detectors, and is defined as:

$$\text{DVI} = \sqrt{(\text{SVI})^2 - (\text{RVI})^2}. \tag{5.4}$$

Exercise 5.2 in this chapter provides a numerical exercise for calculating the quantities SVI, RVI, and DVI.

Definitions for Reference and Field dosimeters

In TLD applications, it is common to divide the available dosimeters in two classes, *reference dosimeters* and *field dosimeters*. The main difference between reference and field dosimeters comes from their different uses. The field dosimeters are to be used for calibrating TLD readers and for monitoring radiation in all dosimetric applications. On the other hand, the reference dosimeters are used to produce an "average response" to which the response of the field dosimeters is normalized.

Typically, the group of reference dosimeters is chosen as a subgroup of the batch of all the available dosimeters. A typical example may be to choose 10 reference dosimeters out of a batch of 100 available dosimeters. It is noted that after an irradiation test, the net TL signal of the 10 reference dosimeters must be closer to the average value than those of the other samples. These 10 reference dosimeters will be considered as representative of the whole batch and should not be used for field applications.

A comprehensive example of how to set up and calibrate a TLD system for radiation monitoring is given by Plato and Miklos [5]. These authors advocated the use of **element correction factors** (ECFs) for personal dosimetry, and presented a 10-step procedure for producing ECFs for a system of 3,000 TLDs. They suggested that the ECF procedure greatly increases the accuracy in measuring doses with TLD systems. The main advantage of using individual ECFs during TL dosimetry is that ECFs represent only variations within the TL elements and not variations due to stability problems with the TLD reader.

The basic idea behind the ECF procedure is to determine the mean response of the *reference* dosimeters, and to generate ECFs for the *field* dosimeters. By applying these individual ECFs, the accuracy of the field dosimeters is greatly improved. It must be noted that ECFs can also be found in the TLD literature under the names "element correction coefficients" (ECCs), or "individual correction factors", or "relative intrinsic sensitivity factors."

A simplified calibration procedure of dosimeters is as follows. The first step consists of annealing all the TLD dosimeters ($i = 1, 2 .. N$) according to the manufacturer's recommended anneal procedure. The dosimeters can then be read out using the appropriate read out cycle to find the intrinsic TL background signal for each dosimeter. This background must be subtracted from the TL signal in all subsequent TL measurements.

The dosimeters are then irradiated to a known dose that will be typical of the application in which the dosimeters will be used. After irradiation, it is best to read out the irradiated dosimeters in a single session if possible, and by using always the same readout cycle. This step determines the TL signal M_i for each dosimeter.

In the next step, one calculates the mean response $\overline{M}$ of the batch of N dosimeters by taking the average

$$\overline{M} = \frac{1}{N} \sum_{i=1}^{N} M_i. \tag{5.5}$$

If desired, the average response of the N dosimeters can be measured several more times using the same anneal and irradiation procedure, in order to improve the accuracy of the measurements.

Once the average response $\overline{M}$ has been determined, one calculates the individual sensitivity factor S_i for the ith dosimeter belonging to the batch of N dosimeters, by using the definition

$$S_i = \frac{\overline{M}}{M_i}, \tag{5.6}$$

where M_i is the net reading of the ith dosimeter.

The individual S_i factors are associated with each i dosimeter and are used to correct the net dosimeter response M_i at any absorbed dose using the expression:

$$M_{(i)cor} = M_i S_i. \qquad (5.7)$$

As shown in numerical Exercise 5.3 in this chapter, the above procedure can improve the reproducibility of the batch of TLD dosimeters and reduce the relative standard deviation from a typical value of 20% down to a few percent.

For a much more detailed description of a recommended calibration procedure for TLD systems, the reader is referred to the work of Plato and Miklos [5].

Exercise 5.1: Lower Detection Limit, D_{LDL}

You are given in the Table 5.1 the readings of 10 TLDs of the same kind. The second column shows the TLD background reading (no irradiation), and column three shows the TLD readings after the TLDs are irradiated with a dose of 3 μGy using a Co^{60} source.

Calculate the lower detection limit, D_{LDL}.

Solution

The *lower detection limit* D_{LDL} is defined as three times the standard deviation of the zero-dose reading, σ_{BKG}, given in units of the absorbed dose (in units of Gy). Thus, we calculate the average and standard deviation of the data in the second column:

$\overline{M}_0$ (average of the zero-dose readings) = 4.24 reader units.

σ_{BKG} (standard deviation of $\overline{M}_0$) = 2.16 reader units.

To obtain D_{LDL}, we need to transform first the reader units into absorbed dose (Gy). For this, we have to calculate the calibration factor of the TL reader Φ_C,

TABLE 5.1. Calculation of the lower detection limit, D_{LDL}

TLD no.	M_0 (reader units) read out after annealing (intrinsic background)	M_i (reader units) read out after irradiation	$M_i - M_{0i}$ readout minus background
1	5.34	799.5	794.16
2	9.91	751.9	741.99
3	3.97	787.2	783.23
4	3.82	877.8	873.98
5	3.99	760.7	756.71
6	4 07	1306.0	1301.93
7	3.26	1000.0	996.74
8	2.99	1003.0	1000.01
9	2.53	935.0	932.47
10	2.55	973.2	970.65
	$\overline{M} \pm \sigma_{BKG} = 4.24 \pm 2.16$		$\overline{M} = 915.19 \pm 168.54$

which is given by the ratio of the calibration dose D_C divided by the average value of the net TL readings $\overline{M}$:

$$\Phi_C = \frac{D_C}{\overline{M}} = \frac{D_C}{\dfrac{1}{N}\displaystyle\sum_{i=1}^{N}(M_i - M_{0i})}, \tag{5.8}$$

where D_C is the calibration dose, N is the number of TLDs, M_i is the reading of the ith TLD, and M_{0i} is the zero-dose reading (background) of the ith TLD.

The average value of the 10 net readings is $\overline{M} = 915.19$ (reader units) with a standard deviation $\sigma = 168.54$ (reader units).

The calibration factor is

$$\Phi_C = \frac{3}{915.19} = 3.3 \text{ } \mu\text{Gy/TL reader unit}$$

and its error is

$$\frac{\Delta \Phi_C}{\Phi_C} = \frac{\Delta D}{D} + \frac{\Delta \overline{M}}{M} = \frac{168.54}{915.19} = 0.18$$

and

$$\Delta \Phi_C = 0.18 \times 3.3 = 0.64 \text{ } \mu\text{Gy/TL reader units}.$$

Finally, we have

$$\Phi_C = 3.3 \pm 0.6 \text{ } \mu\text{Gy/TL units}$$

and

$$D_{\text{LDL}} = (6.48) \times (3.3) = 21 \text{ } \mu\text{Gy}.$$

Exercise 5.2: Reproducibility Measurements

You are given in Table 5.2 the readings of 10 TLDs. Each TLD was annealed using the appropriate annealing procedure, then irradiated with the same dose, and the TL signal was read out. This procedure has been carried out over five cycles.

(1) Study the variability of the TL system. This represents a measure of the variability of the combined TL system consisting of the TL reader, the annealing, and the irradiation procedures, and the TL dosimeters.
(2) Study the variability of the TL reader only.
(3) Study the variability of the TLDs only.

Solution

(1) By working across each row of Table 5.2, we calculate the average, the standard deviation, and the covariance for each detector, with the results of these calculations shown in Table 5.3. By using the last column of Table 5.3, we also find the average covariance $\overline{CV}$ of the group of dosimeters.

TABLE 5.2. Data table for reproducibility of TLD measurements

Dosimeter no.	1st reading (TL reader units)	2nd reading (TL reader units)	3rd reading (TL reader units)	4th reading (TL reader units)	5th reading (TL reader units)
1	6.957	6.836	6.702	6.735	6.898
2	6.752	7.065	6.804	6.665	6.956
3	6.686	6.764	6.588	6.756	6.630
4	6.708	6.532	6.606	6.830	6.826
5	6.853	6.833	6.980	6.860	6.978
6	6.731	6.783	6.782	6.819	6.852
7	6.759	6.578	6.629	6.672	6.654
8	6.686	6.836	6.944	6.840	6.762
9	6.843	6.662	6.772	6.797	6.637
10	6.788	6.696	6.557	6.627	6.474

This $\overline{CV}$ value expressed in % gives us the SVI of the TL system.

$$\overline{CV} = 0.015.$$
$$\%\overline{CV} = SVI = 1.5\%.$$

As discussed in the introductory sections of this chapter, the quantity $\% \overline{CV} = SVI$ gives a measure of the reproducibility of the whole system.

(2) By working across each column of Table 5.2, we calculate the mean readings for each cycle, the average of the mean readings, their standard deviation, and the associated covariances, with the results of these calculations shown in Tables 5.3 and 5.4. These covariances give us the variability index of the TL reader only, or RVI.

$$SD = 0.015$$
$$\%CV = RVI = SD/mean = 0.015/6.760 = 0.0022 = 0.2\%.$$

As discussed in the introductory sections of this chapter, the quantity $\%CV = RVI$ gives a measure of the long-term reproducibility of the reader system.

TABLE 5.3. Statistical analysis of data in Table 5.2

Dosimeter no.	Mean values of the five readings	Standard deviation	CV
1	6.786	0.146	0.022
2	6.848	0.161	0.024
3	6.685	0.077	0.012
4	6.700	0.132	0.020
5	6.901	0.072	0.010
6	6.793	0.045	0.007
7	6.658	0.066	0.010
8	6.814	0.096	0.014
9	6.742	0.089	0.013
10	6.628	0.121	0.018
$\overline{CV}$			0.015
$\%\overline{CV}$			1.5%

TABLE 5.4. Calculation of reader variability index (RVI), system variability index (SVI), and variability index of the detectors (DVI)

1st reading	2nd reading	3rd reading	4th reading	5th reading	mean	SD	%CV
6.776	6.759	6.736	6.760	6.767	6.760	0.015	0.2%

(3) The DVI is then given by the following expression:

$$DVI = \sqrt{(SVI)^2 - (RVI)^2} = \sqrt{2.25 - 0.04} = 1.5\%.$$

From the results obtained, it is evident that a better reproducibility can be achieved by carefully selecting the individual dosimeters or by using the individual sensitivity factors associated to each dosimeter, as discussed in the next exercise.

Exercise 5.3: Individual Correction Factors, S_i

In Table 5.5, the net response of a group of nine TLDs at a specific dose of 3.78 µGy are given. The aim of the exercise is to study the effect of the individual correction factors S_i on the standard deviation of the data. Calculate the individual correction factors for each of the nine TLDs and the standard deviation of the data for each dosimeter before and after using the S_i corrections.

Solution

The general definition of the individual sensitivity factor, S_i, where i stands for the ith dosimeter belonging to a batch of N TLDs is

$$S_i = \frac{\overline{M}}{M_i}, \tag{5.9}$$

TABLE 5.5. Data table for the individual sensitivity factor, S_i exercise

Dose (µGy)	Net readings (counts)
3.78	12,993
3.78	11,739
3.78	8,212
3.78	8,339
3.78	8,527
3.78	9,098
3.78	7,571
3.78	9,948
3.78	9,749

TABLE 5.6. Calculation of the individual sensitivity factor, S_i

Dose (µGy) M_i	Net readings counts M_i	Average (counts) $\overline{M}$	SD (%) before correction	S_i factor $S_i = \overline{M}/M_i$	Corrected readings $M_{(i)cor} = M_i S_i$	Corrected average	SD(%) after correction
3.78	12,993	9,575	18.6	0.737	9,575.84	9,575.66	0.03
	11,739			0.816	9,579.02		
	8,212			1.166	9,575.19		
	8,339			1.148	9,573.17		
	8,527			1.123	9,575.82		
	9,098			1.052	9,571.10		
	7,571			1.265	9,577.32		
	9,948			0.963	9,579.92		
	9,749			0.982	9,573.52		

where M_i is the net reading of the ith dosimeter and $\overline{M}$ is the average of the net readings of the N dosimeters.

As discussed in the introductory sections of this chapter, the individual S_i factors are associated with each i-dosimeter and are used to correct the net dosimeter response M_i at any absorbed dose

$$M_{(i)cor} = M_i S_i. \qquad (5.10)$$

The third column in Table 5.6 contains the average of the net readings for the group of dosimeters, while the fourth column contains the percent standard deviation (SD%) for the nine readings.

The fifth column is calculated using the definition of S_i and the sixth column shows the corrected values of the net readings. The last column in Table 5.6 shows the % SD when using the corrected net readings.

As can be seen from the Table, by using the individual sensitivity factor S_i the % standard deviations are reduced from 18% to only 3%.

Exercise 5.4: Relative Standard Deviation Versus Dose

The dosimetric properties of a TLD system are given by the combination of the following factors:

- TL reader (PM dark current, readout profile)
- Thermoluminescent material (batch quality, batch history)
- Annealing procedure
- Calibration (irradiation system)
- Zero-dose reading

All the previous factors affect the reproducibility of the TLD system. Furthermore, the reproducibility results are also dependant on the irradiation dose.

As discussed in the introduction to this chapter, the reproducibility of a TLD system can be expressed as the variation of the standard deviation $s(D)$ of the TL

TABLE 5.7. TL readings and their
experimental relative standard
deviations

Dose (μGy)	Experimentals SD%
3.78	18.6
4.12	13.8
6.14	11.9
11.2	8.3
24.5	4.5
49.1	2.3
147	2.1
294	2.4
589	1.5
1,178	1.5

readings at different doses D. This variation can often be expressed by using a two-parameter fit [1]:

$$s(D) = \left[\frac{A^2}{D^2} + B^2 \right]^{1/2}, \qquad (5.11)$$

where A is absolute SD at very low dose (zero dose), expressed in dose units, B is relative SD at high dose expressed in reader units.

In this exercise, a typical example of the variation of $s(D)$ with the dose D is given.

The first two columns of Table 5.7 list the experimental relative standard deviation $s(D)$ at each dose D. Find the coefficients A, B such that the theoretical plot from equation (5.11) fits the experimental data.

Solution

By inspecting the second column of Table 5.7, we see that the standard deviation at large doses becomes constant and equal to 1.5%. This is in agreement with equation (5.11), which tells us that the value of the constant S(D) at large doses should be constant and equal to $B = 1.5$.

Equation (5.11) tells us that by graphing the quantities $s(D)^2 - B^2$ versus $1/D^2$, we should find a linear function with a slope equal to A^2, as shown in Figure 5.2.

The values of the constants A and B are therefore equal to $A^2 = \text{slope} = 4180.39$ and $B^2 = 1.25$. Figure 5.3 shows that the experimental and theoretical data are in a reasonably good agreement.

Exercise 5.5: Dose Monitoring in a Nuclear Medicine Department

In the nuclear medicine department of a hospital, it was decided to monitor the radiation level. In order to perform this measurement, a batch of TLDs was annealed

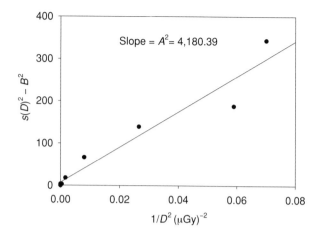

FIGURE 5.2. The standard deviations $s(D)^2 - B^2$ versus $1/D^2$ fit a linear function with slope A^2.

and the main dosimetric characteristics were studied to determine the fading factor of the material (λ).

Subsequently, a dose monitoring experiment was carried out over a period of 2 months by placing several TLDs in various positions in the hospital. The environmental dose value corresponding to the TL emission recorded at the end of the storage period is incorrect, and needs to be corrected for the fading effect of the TLDs.

Find the corrected environmental dose over a period of 2 months (1,440 hours) when the fading factor of the TL material is $\lambda = 3.2 \times 10^{-4}$ h^{-1} and the measured incorrect environmental dose at the end of the 2 months is $D_{BF}(t_S) = 42.47$ mGy.

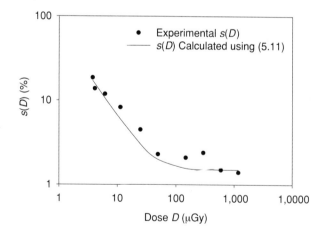

FIGURE 5.3. Comparison of the experimental values of $s(D)$ and the theoretical relative standard deviations $s(D)$ calculated from equation (5.11).

Solution

Let us indicate by $\dot{D}_B$ the corrected environmental dose rate.

During the monitoring time t_s two effects are in competition: the first is the growth of the TL information due to the environmental dose, and the second is the fading of the TL signal in the TLDs. The equation describing this competing process is as follows [6, p. 138]:

$$D_{BF}(t_s) = \frac{\dot{D}_B}{\lambda}\left[1 - \exp(-\lambda\, t_s)\right] \qquad (5.12)$$

from which the corrected environmental dose rate $\dot{D}_B$ can be determined:

$$\dot{D}_B = \frac{D_{BF}(t_s)\,\lambda}{1 - \exp(-\lambda\, t_s)}. \qquad (5.13)$$

By substituting the given numerical values in equation (5.13), we find

$$\dot{D}_B = \frac{D_{BF}(t_s)\,\lambda}{1 - \exp(-\lambda\, t_s)} = \frac{42.47 \times 3.2 \times 10^{-4}}{1 - \exp(-3.2 \times 10^{-4} \times 1,440)} = 0.037\ \text{mGy h}^{-1}$$

and the corresponding corrected environmental dose over the period of storage is equal to $0.037\ \text{mGy h}^{-1} \times 1,440\ \text{hours} = 53\ \text{mGy}$.

Exercise 5.6: Determination of the Self-Dose in a TL Material

During a study of the dosimetric characteristics of a thermoluminescent phosphor, it was found that it was affected by a self-dose effect, due to the presence of isotopic substances in the material. In order to take into account this effect and to correct the TL emission in dosimetric applications, the thermoluminescent material was annealed and then stored in a lead box to avoid any external irradiation. The period of storage was 1,440 hours and at the end of this time, the TL was measured to determine the self-dose, which was found equal to $D_{\text{SDF}}(t_s) = 10.54\ \mu\text{Gy}$.

Calculate the corrected value of the self-dose when the fading factor of the material λ is known to be $\lambda = 5.2 \times 10^{-4}\ \text{h}^{-1}$.

Solution

The equation describing the corrected self-dose rate is [6, p. 138]

$$\dot{D}_{SD} = \frac{D_{SDF}(t_s)\lambda}{1 - \exp(-\lambda t_s)}. \qquad (5.14)$$

And the corresponding corrected self-dose will be equal to

$$D_{SD} = \dot{D}_{SDF} \times \text{time} = \frac{D_{SDF}(t_s)\lambda}{1 - \exp(-\lambda t_s)} \times \text{time}. \qquad (5.15)$$

By substituting the given numerical values in equation (5.15), we find

$$D_{SD} = \frac{D_{SDF}(t_s)\lambda}{1 - \exp(-\lambda t_s)} \times \text{time} = \frac{10.54 \times 5.2 \times 10^{-4}}{1 - \exp(-5.2 \times 10^{-4} \times 1,440)} 1,440$$

$$= 0.0104 \times 1,440 \ \mu\text{Gy}.$$

The corrected total self-dose is therefore 14.97 μGy.

Exercise 5.7: Simulation of Thermal Quenching in TL Materials

The phenomenon of thermal quenching is present in several important thermoluminescence materials, such as quartz and Al_2O_3. The purpose of this exercise is to simulate the influence of the thermal quenching effect on the experimentally measured first-order TL glow curves by using the following procedure.

(a) Evaluate the first-order TL peak intensity with parameters $E = 1$ eV, $s = 10^{12}$ s^{-1} and $n_0 = 10^3$ cm^{-3} when no thermal quenching is present, and also when thermal quenching is present. Use the thermal quenching parameters $W = 0.85$ eV and $C = 10^{11}$ s^{-1}.

(b) Find the dependence of the peak integrals (area under the glow curve), peak maximum temperature, and FWHM on the heating rate β used during the measurement of the TL glow curve.

(c) Evaluate the thermal quenching parameters W and C by using the experimentally measured quenched TL intensity.

(d) Evaluate the influence of thermal quenching on the calculated activation energy E by applying the peak shape methods and the variable heating rate method of analysis.

Solution

(a) A first-order glow peak is described by the usual Randall–Wilkins expression (see equation 1.4 in introduction):

$$I(T) = n_0 s \exp\left(-\frac{E}{kT}\right) \exp\left[-\frac{s}{\beta} \int_{T_0}^{T} \exp\left(-\frac{E}{kT'}\right) dT'\right]. \tag{1.4}$$

As discussed in Chapter 1, the integral appearing in equation (1.4) can be evaluated by a series of approximation, and the TL intensity can be written in the approximate form:

$$I(T) = n_0 s \exp\left(-\frac{E}{kT}\right) \exp\left[-\frac{skT^2}{\beta E} \exp\left(-\frac{E}{kT}\right)\left(1 - \frac{2kT}{E}\right)\right]. \tag{5.16}$$

The thermal quenching efficiency $\eta(T)$ is given by the expression [7]

$$\eta(T) = \frac{1}{1 + C \exp\left(-\dfrac{W}{kT}\right)}, \tag{5.17}$$

where C and W are the thermal quenching pre-exponential factor and activation energy, respectively.

The experimentally observed TL glow curve corresponds to a quenched TL glow peak with intensity denoted by $I_{QU}(T)$. This quenched TL intensity is found by multiplying the quenching efficiency $\eta(T)$ by the unquenched TL intensity $I_{UNQ}(T)$, i.e.

$$I_{QU}(T) = I_{UNQ}(T)\eta(T). \tag{5.18}$$

The values of all parameters given in this exercise are such that the influence of thermal quenching is negligible at the lowest available heating rate. This type of calculation is easily setup on a spreadsheet that contains the temperature T, the calculated unquenched TL intensity using equation (5.16), and the quenched TL intensity calculated using equation (5.18).

A series of simulated unquenched and quenched glow peaks are calculated in this manner, and they are shown in Figure 5.4.

From the calculated TL glow curves of Figure 5.4, one can evaluate easily the peak integral (represented by the area under the curve), the temperature of maximum TL intensity T_M, the FWHM, and the symmetry factor μ_g of each glow peak. Special care must be taken during the peak integral (area) evaluation. The peak integral is evaluated by summing the TL intensity from a temperature T_0 up to the temperature at which the glow peak ends.

However, two more actions are necessary: (i) To multiply by the temperature interval ΔT between two successive TL intensities and (ii) To divide by the heating rate β. The expression for peak integral evaluation is

$$\text{Peak integral} = \frac{1}{\beta} \cdot \sum_i I_{QU}(T) \cdot \Delta T_i. \tag{5.19}$$

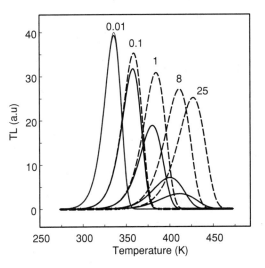

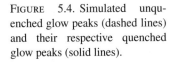

FIGURE 5.4. Simulated unquenched glow peaks (dashed lines) and their respective quenched glow peaks (solid lines).

TABLE 5.8. The temperature of maximum TL intensity T_M, the FWHM, and the geometrical shape factor μ for both quenched and unquenched TL glow curves

β (K s^{-1})	Area	T_{MUNQ}[K]	T_{MQU}[K]	FW_{UNQ}	FW_{QU}	μ_g	E(eV)
0.01	987.374	336.293	336.16	22.831	22.846	0.4237	1.032
0.05	946.311	351.793	351.269	24.942	25.017	0.4248	1.031
0.1	909.029	358.902	357.999	25.942	26.096	0.4257	1.029
0.5	746.875	376.533	373.734	28.500	29.310	0.4301	1.012
1	643.664	384.655	380.419	29.718	31.247	0.4335	0.995
2	529.153	393.123	387.023	31.014	33.708	0.4374	0.966
4	414.312	401.96	393.585	32.394	36.814	0.4414	0.925
8	309.511	411.19	400.183	33.867	40.665	0.4446	0.871
12	255.764	416.782	404.101	34.774	43.290	0.4459	0.835
18	208.504	422.523	408.091	35.718	46.199	0.4464	0.796
25	175.088	427.287	411.396	36.510	48.769	0.4462	0.763
35	145.27	432.276	414.863	37.348	51.600	0.4455	0.727

Column 2 in Table 5.8 shows the results of calculating the peak integral for the quenched TL glow curves shown in Figure 5.4. The behavior of the peak integral as a function of the heating rate β is also shown in graphical form in Figure 5.5. The unquenched integral is expected to remain constant and equal to $A = 10^3$ and to be independent of the heating rate, whereas the quenched peak integral drastically decreases as the heating rate increases.

The temperature of maximum TL intensity T_M and the FWHM for both quenched and unquenched TL glow curves are also shown in Table 5.8 together with the geometrical shape factor μ for the quenched glow curves. The value of μ for the unquenched curves is of course constant and equal to the first-order kinetics value of $\mu = 0.42$.

The behavior of the temperature of maximum TL intensity peak maxima (T_{MUNQ} and T_{MQU} in columns 3 and 4 in Table 5.8) is shown as a function of the heating rate in Figure 5.6. It is clear that as the heating rate increases, the peak maximum

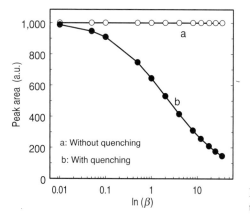

FIGURE 5.5. The behavior of peak integral as a function of the heating rate.

FIGURE 5.6. The behavior of peak maxima as a function of the heating rate.

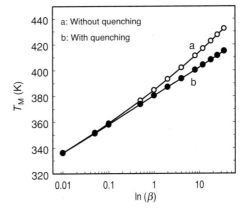

T_M of the quenched glow peak, as well as the glow curve as a whole, shift to lower temperatures due to the effect of thermal quenching.

The behavior of the FWHM (columns 5 and 6 in Table 5.8) as a function of the heating rate is shown in Figure 5.7. It is interesting to note that as the heating rate increases the glow peak becomes much broader due to thermal quenching. On the other hand, a slight increase of the symmetry factor (column 7 in Table 5.8) is also caused by the thermal quenching phenomenon.

(b) It is possible to evaluate the thermal quenching parameters C and W from a set of quenched experimental TL data. If I_{QUE} is the quenched peak integral and A is the constant unquenched peak integral, then according to equations (5.18) and (5.19), we have

$$I_{QUE} = A\eta(T). \qquad (5.20)$$

Since in practice most TL peaks cover a narrow range of temperatures, we can approximate the quenching function $\eta(T)$ by its value at the peak maximum $\eta(T_M)$.

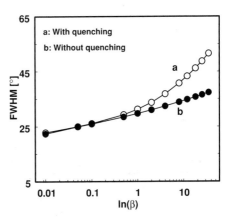

FIGURE 5.7. Behavior of FWHM as a function of heating rate.

In this case, equation (5.20) is written as

$$I_{QUE} = \frac{A}{1 + C \exp\left(-\dfrac{W}{kT_M}\right)}. \tag{5.21}$$

Equation (5.21) can also be rearranged as

$$\frac{A}{I_{QUE}} - 1 = C \exp\left(-\frac{W}{kT_M}\right). \tag{5.22}$$

In some cases the thermal quenching effects are present even from the lowest available heating rate β. In these cases, the value of constant A that is equal to the peak integral of the unquenched glow peak is not known. In such situations, the thermal quenching parameters can be evaluated by a fitting of the experimental peak integral to equation (5.21), with A, C, and W being the free fitting parameters.

In most practical cases, however, the influence of thermal quenching manifests itself for higher heating rates, whereas there is usually only a small thermal quenching effect for lower heating rates. Therefore, the peak integral of the unquenched glow peak is known and is equal to the constant A. Equation (5.22) now tells us that a plot of $\ln(A/I_{QUE} - 1)$ versus $1/kT_M$ will yield a straight line with slope $-W$ and intercept $\ln(C)$, from which C can be evaluated.

By using the values of I_{QU} and TM_{QU} from columns 2 and 4 in Table 5.8, and by using the value of the constant peak integral $A = 1,000$, the plot of $\ln (A/I_{QUE} - 1)$ versus $1/kT_M$ is shown in Figure 5.8.

The quenching parameters obtained from Figure 5.8 are $W = 0.9255 \pm 0.0045$ eV and $C = 1.019 \times 10^{12}$ s^{-1}. The value of W obtained differs by almost 8% from the value of $W = 0.85$ eV used to produce the simulated data, whereas the value of C is larger by an order of magnitude (10^{11} s^{-1}). The reason for these differences is that the method applied here is only approximate, since it

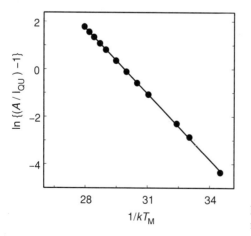

FIGURE 5.8. Plot for quenching parameters evaluation.

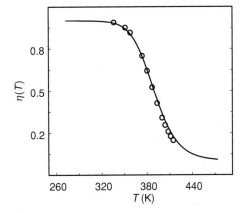

involves a single value $\eta(T_M)$ as representative of the thermal quenching factor $\eta(T)$ across the whole glow peak.

However, the differences between the calculated and the original quenching parameters are not as significant as they may seem. This can be seen from the evaluation of $\eta(T)$ using both the original parameters C and W, as well as the values of C and W obtained by the present approximate method. The results are shown in Figure 5.9.

The solid line corresponds to the values of $\eta(T)$ obtained by using the original W and C values (0.85 eV, 10^{11} s^{-1}), whereas the open circles are calculated using the W and C values obtained by the present approximate method (0.925 eV, 1.019×10^{12} s^{-1}). It is clear that both sets of W and C values give essentially the same $\eta(T)$ graphs for heating rates up to 8 K/s, with an accuracy 5% or better.

A generalization of the present method for the important dosimetric material Al_2O_3 can be found in reference [7].

(c) As discussed in Chapter 1, all peak shape methods are based on the values of the temperature of maximum TL intensity T_M and on the value of the FWHM of the glow peak. Taking into account the results of columns 4 (T_{MQU}) and 5 (FWHM) of Table 5.8 and using the Chen peak shape methods for general-order kinetics [8], the influence of the thermal quenching on the activation energy E evaluation can be estimated. We note that it is necessary to use the general-order equation for the activation energy E because of the variation of the symmetry factor with the heating rate (column 7 in Table 5.8).

The results are shown in the last column of Table 5.8, where we can see the dramatic effect of the thermal quenching phenomenon on the values of E calculated using the peak shape methods.

Using the simulated results in Table 5.8, the influence of the thermal quenching on the variable heating rate method of finding the activation energy E can also be evaluated. The variable heating rate method of analysis consists of a plot of $\ln(T_M^2/\beta)$ versus $1/kT_M$, which is a straight line with slope equal to E and intercept equal to $\ln(Ek/s)$. Using the data of columns 1 and 4 in Table 5.8, the E and s

values obtained from the plot of $\ln(T_M^2/\beta)$ versus $1/kT_M$ is $E = 1.196 \pm 0.02$ eV and $s = 8.085 \times 10^{14}$ s^{-1}.

The conclusion is that when thermal quenching is present, the variable heating rate method overestimates significantly the values of E and s relative to the original values of E and s used in the simulation ($E = 1$ eV, $s = 10^{12}$ s^{-1}).

Exercise 5.8: The Effect of Thermal Quenching on the Initial Rise Method of Analysis

The purpose of this exercise is to study the effect of thermal quenching on the initial rise method of analysis. A series of quenched TL glow curves are simulated and the initial rise method of analysis is applied in order to find the activation energy E.

A first-order glow peak with parameters $E = 1.1$ eV, $n_0 = 10^3$ m^{-3} and $s = 10^{12}$ s^{-1} is influenced by thermal quenching described by the parameters $W = 0.85$ eV and $C = 10^{11}$ s^{-1}.

(a) Evaluate the quenching efficiency $\eta(T)$, the unquenched glow peak $I_{UNQ}(T)$ and the quenched glow peak $I_{QU}(T)$ for heating rates $\beta = 0.01, 1, 5, 10$, and 20 K s^{-1} as in the previous exercise.

(b) Plot $\eta(T), I_{UNQ}(T), I_{QU}(T)$ and evaluate the activation energy E_{QU} by applying the initial rise method to the quenched TL intensity $I_{QU}(T)$. Discuss how thermal quenching affects the initial rise method for evaluating the activation energy.

(c) Apply a suitable correction method to evaluate the real activation energy from the activation energy E_{QU} obtained from the quenched glow peaks.

Solution

(a)As discussed in the previous exercise, the simulated TL glow curves can be calculated using the following approximate expression in a spreadsheet:

$$I(T) = n_0\, s \exp\left(-\frac{E}{kT}\right) \exp\left[-\frac{skT^2}{\beta E} \exp\left(-\frac{E}{kT}\right)\left(1 - \frac{2kT}{E}\right)\right]. \quad (5.23)$$

The thermal quenching efficiency $\eta(T)$ is given by the expression [7]

$$\eta(T) = \frac{1}{1 + C \exp\left(-\dfrac{W}{kT}\right)}, \quad (5.24)$$

where C and W are the thermal quenching pre-exponential factor and activation energy, respectively. The quenched glow peak $I_{QU}(T)$ is calculated by using the expression

$$I_{QU}(T) = I_{UNQ}(T)\eta(T). \quad (5.25)$$

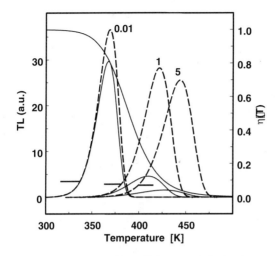

FIGURE 5.10. Right axis: Thermal quenching efficiency $\eta(T)$. Left axis: unquenched (dashed lines) and quenched (solid lines) glow peaks for heating rates 0.01, 1, and 5 K s^{-1}. The horizontal solid lines correspond to 10% of the unquenched peak maximum intensity, I_M.

A series of simulated unquenched and quenched glow peaks calculated in the same method as in the previous exercise are shown in Figure 5.10 for heating rates of 0.01, 1, and 5 K s^{-1}.

Before attempting to apply the initial rise method of analysis, one has to define the temperature region in which the method is going to be applied. As a practical rule, the TL intensity at the upper temperature limit should not exceed approximately 10% of the maximum intensity I_M. The horizontal lines on the unquenched glow peaks in Figure 5.10 show the exact location of this 10% limit.

The location of the lower temperature limit is not so obvious because the simulation starts from 273 K, and the TL is evaluated even at temperatures far from the temperature region of the glow peak. The simplest assumption would be to take the lower temperature of the initial rise region as located 20–30 K below the upper temperature limit.

The results of Figure 5.10 indicate that for the glow peak measured with a heating rate of 0.01 K s^{-1} the initial rise region is located in a temperature area where the values of the quenching function $\eta(T)$ are very close to 1. Therefore, according to equation (5.25), the TL in this temperature region is influenced very little by thermal quenching. On the other hand, inspection of Figure 1 shows that the rest of the TL glow peaks at heating rates 1 and 5 K s^{-1} are strongly influenced by thermal quenching effects.

On general grounds, one would expect that the initial rise method for activation energy evaluation will be much less influenced by the thermal quenching than the peak shape methods and variable heating rate methods of analysis. This can be expected because the latter methods depend on T_M and the FWHM, which are highly influenced by the thermal quenching as shown in the previous exercise.

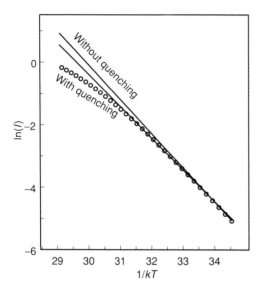

FIGURE 5.11. Initial rise plots with
5 K s^{-1}.

(b) The initial rise method was applied to the simulated quenched glow peaks and the activation energy, E_{QU}, is obtained. An example of an initial rise plot for $\beta = 5$ K s^{-1} is shown in Figure 5.11. The upper straight line corresponds to an initial rise plot of the unquenched glow peak. The open circles correspond to the initial rise plot of the quenched glow peak in the same temperature region.

The middle solid line is the initial rise line of the quenched glow peak extrapolated to the whole temperature region. The slope of the middle solid line represents E_{QU}, and it is seen to be lower than the corresponding slope of the unquenched data that gives a value of $E = 1.0945$ eV, very close to the original E value of $E = 1.1$ eV used in the simulation.

From Figure 5.11, it is clear that the linearity of the initial rise plot is affected by the thermal quenching effect and that this effect is smaller at lower temperatures.

The resulting values of the activation energy using the quenched data are shown in Table 5.9 column 2, with the original value being $E = 1.1$ eV. The difference $\Delta E = 1.1 - E_{QU}$ is shown in column 3. The results in the Table show that the E-values obtained using the initial rise method on quenched TL data are underestimating the real activation energy E. In the present example, the underestimation is of the order of 10% for the highest available heating rates.

(c) According to equation (5.25), the initial rise method is simulated by an expression of the form:

$$I_{QU}(T) = n_0 s \, \exp\left[-\frac{E_R}{kT}\right] \cdot \eta(T), \qquad (5.26)$$

where E_R is the real value of the activation energy.

The experimental initial rise region for the quenched data is expressed by

$$I_{QU}(T) = n_0 \, s_{QU} \, \exp\left[-\frac{E_{QU}}{kT}\right], \qquad (5.27)$$

where, S_{QU} and E_{QU} are the frequency factor and the activation energy of the quenched experimental glow peak, respectively.

Taking the natural logarithm of equations (5.26) and (5.27), we have

$$\ln[I_{QU}(T)] = \ln[n_0 s] - \frac{E_R}{kT} + \ln[\eta(T)] \qquad (5.28)$$

$$\ln[I_{QU}(T)] = \ln[n_0 s_{QU}] - \frac{E_{QU}}{kT}. \qquad (5.29)$$

Taking the derivatives with respect to $1/kT$, we have

$$\frac{d\ln[I_{QU}(T)]}{d[1/kT]} = -E_R + \frac{d\ln[\eta(T)]}{d[1/kT]} \qquad (5.30)$$

$$\frac{d\ln[(I_{QU}(T))]}{d[1/kT]} = -E_{QU}. \qquad (5.31)$$

From equations (5.30) and (5.31) we have

$$E_R = E_{QU} + \frac{d\ln[\eta(T)]}{d[1/kT]}. \qquad (5.32)$$

Taking into account equation (5.24), the difference $\Delta E = E_R - E_{QU}$ is

$$\Delta E = \frac{WC \, \exp(-W/kT)}{1 + C \, \exp(-W/kT)}. \qquad (5.33)$$

Equation (5.33) was derived by Petrov and Bailiff [9] and expresses the correction term to be applied to the activation energy E when thermal quenching is present. In applying equation (5.33) one has to decide what to use for the temperature T. Petrov and Bailiff [9] suggested that this temperature can be the one corresponding to the middle of the initial rise temperature region, provided that this region is not broader than 10–15°C.

The results of this correction procedure are given in columns 3, 4, and 5 of Table 5.9. Column 3 contains the difference between $E_R = 1.1$ eV and E_{QU} (column 2). Column 4 shows the temperature region in which the initial rise plot was performed in order to obtain the values of E_{QU} of column 2. Column 5 gives

TABLE 5.9. Calculation of the activation energy E using the quenched data

β(K s^{-1})	E_{QU} (eV)	ΔE (eV)	IR region (K)	ΔE [8] (eV)	E_R (eV)
0.01	1.085	0.015	311–337	0.009	1.0912
1	1.055	0.045	333–365	0.043	1.0985
5	1.018	0.082	345–369	0.077	1.0993
10	1.000	0.100	348–372	0.095	1.0995
20	0994	0.106	349–373	0.102	1.0997

the difference $\Delta E = E_R - E_{QU}$ evaluated by using the equation (5.33). The agreement between columns 3 and 5 is satisfactory.

An alternative method to find the E_R from the quenched initial rise data was suggested by Kitis [10]. Equation (5.26) can be written as

$$\ln\left[\frac{I_{QU}(T)}{\eta(T)}\right] = \ln(n_0 s) - \frac{E_R}{kT}. \tag{5.34}$$

By substituting the value of the thermal quenching factor $\eta(T)$ from equation (5.24), equation (5.34) is transformed into

$$\ln[I_{QU}(1 + C \cdot \exp(-W/kT))] = \ln(n_0 s) - \frac{E_R}{kT}. \tag{5.35}$$

By applying equation (5.35) to the quenched initial rise region of the glow peak, one obtains directly the real activation energy E_R. The results of this procedure are listed in column 6 of Table 9. The values of E_R shown in column 6 are in excellent agreement with the original value $E_R = 1.1$ eV used for the simulations. The advantage of this procedure relative to that by Petrov and Bailiff [9] is that the values of $\eta(T)$ are taken into account in the whole temperature region of the initial rise plot, instead of using only one value $\eta(T_M)$ corresponding to the temperature in the middle of the initial rise region.

Exercise 5.9: TL-Like Presentation of Phosphorescence Decay Curves

Simulate the phosphorescence decay of a first-order glow peak with parameters $I_0 = 10^3$, $E = 1$ eV, and $s = 10^{12}$ s^{-1} at the decay temperatures 350, 355, 360, 365, 370, and 375 K.

(a) Compare the phosphorescence decay curves with the so-called TL-like presentation form of phosphorescence for both first-order and general-order kinetics.

(b) Evaluate the trapping parameters E and s by using the simulated data for the TL-like presentation of phosphorescence.

(c) Compare the phosphorescence decay curves with the TL-like presentation curves for different kinetic orders b between $b = 1$ and $b = 2$. Indicate the similarities and differences between the usual TL glow peaks and the TL-like presentation of the data. Evaluate the peak shape characteristics of the TL-like presentation decay curves.

(d) Investigate whether thermal quenching influences the phosphorescence decay of the material at high temperatures. Also investigate the effect of quenching on the evaluation of the trapping parameters E and s by using phosphorescence data. Use the following thermal quenching parameters: $W = 0.85$ eV and $C = 10^{11}$ s^{-1}.

Solution

(a) The intensity of phosphorescence decay as a function of time and at a given decay temperature T_d is given for first-order kinetics by the expression

$$I(t) = I_0 \exp\left(-\frac{t}{\tau}\right) \tag{5.36}$$

and for general-order kinetics by the expression

$$I(t) = I_0 \left\{1 + (b-1)\frac{t}{\tau}\right\}^{-\frac{b}{b-1}} \tag{5.37}$$

with $\tau = s_{\text{eff}} \exp(-E/k\,T_d)$, and I_0 represents the phosphorescence intensity at time $t = 0$. The "effective frequency factor" s_{eff} is equal to $s_{\text{eff}} = s$ for first-order kinetics, and $s_{\text{eff}} = s'' = s'n_0^{b-1}$ for general order kinetics.

By defining $x = \ln(t)$ and $t = \exp(x)$, equation (5.36) is rewritten as ([11]–[13])

$$I(t) = I_0 \exp\left(-\frac{\exp(x)}{\tau}\right). \tag{5.38}$$

Multiplying both sides by t, and using a new variable $y = It$, one gets from equation (5.36) the so-called *TL-like presentation form of the first-order phosphorescence decay curve*, i.e.

$$y = I_0 \exp(x) \cdot \exp\left(-\frac{\exp(x)}{\tau}\right). \tag{5.39}$$

Working in a similar way with equation (5.37), one gets the *TL-like presentation form of the general-order phosphorescence decay curve*, i.e.

$$y = I_0 \exp(x) \left\{1 + (b-1)\frac{\exp(x)}{\tau}\right\}^{-\frac{b}{b-1}}. \tag{5.40}$$

Equations (5.39) and (5.40) describe peak-shaped curves mathematically similar to those of the TL intensity as a function of temperature, so they were termed *TL-like presentation of phosphorescence decay*. The condition for the maximum intensity in the TL-like presentation of TL can be found by solving the equation $dy/dx = 0$. The result for both first- and general- order kinetics is

$$x_m = \ln(\tau) \tag{5.41}$$

from which the following is derived:

$$\tau = \exp(x_m) = s_{\text{eff}} \cdot \exp\left(-\frac{E}{kT_d}\right). \tag{5.42}$$

Equation (5.42) shows that the constant $\tau(T_d)$ can be evaluated directly from the TL-like presentation curve, regardless of the order of kinetics. Moreover, by recording the phosphorescence at two or more decay temperatures, the kinetic parameters E and s_{eff} can be computed from equation (5.42) by the relation

$$x_m(T_d) = \frac{E}{kT_d} - \ln(s_{eff}). \qquad (5.43)$$

From an experimental point of view, one measures the phosphorescence at two decay temperatures T_1 and T_2. The values of $x_m(T_1)$ and $x_m(T_2)$ are obtained directly from the maxima of the peak-shaped curves and by using equation (5.41) with two different temperatures, the activation energy is computed by the relation

$$E = kD\frac{T_1 T_2}{T_1 - T_2} \qquad (5.44)$$

with

$$D = x_m(T_2) - x_m(T_1). \qquad (5.45)$$

By using a spreadsheet program, one can easily simulate phosphorescence curves as a function of time, as well as the corresponding TL-like presentation of phosphorescence decay curves as follows:

(i) Evaluate the decay curve using equation (5.36) or (5.37) for the respective kinetic order and obtain two columns, with time (t) and phosphorescence intensity (I).
(ii) Evaluate the columns $x = \ln(t)$ and $y = It$.
(iii) Plot y versus x. This type of calculation can easily be set up in a spreadsheet.

Some specific examples calculated for the given trapping parameters and decay temperatures are shown in Figure 5.12. In Figure 5.12(a), one can easily see that the exponential decay curve appears featureless when compared with the peak-shaped TL-like presentation. In the case of the exponential form, one must plot the data in a semilog scale and evaluate the time constant τ by a least squares fit, whereas in the TL-like presentation the time constant τ is directly obtained by

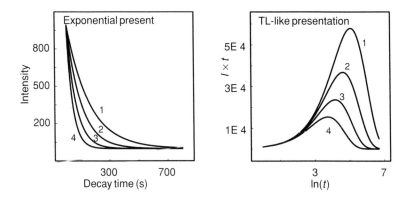

FIGURE 5.12. (a) Shape of phosphorescence decay curves as a function of time. (b) TL-like presentation of the phosphorescence decay curves in (a). The decay temperatures used are (1) 350 K, (2) 355 K, (3) 360 K, and (4) 365 K.

TABLE 5.10. The values of x_m
obtained from the maximum of the
TL-like presentation decay curves

$T_d[K]$	x_m
350	5.52545
355	5.05625
360	4.60517
365	4.15906
370	3.73767
375	3.33220

the location of the maximum of the peak-shaped curves in Figure 5.12(b). As the decay temperature increases, the maximum clearly shifts to lower $\ln(t)$ values, i.e. the time constant τ decreases.

(b) By simulating the phosphorescence decay curves for the decay temperatures given, the values of x_m are directly obtained from the maximum of the TL-like presentation decay curves. The results are shown in Table 5.10. Using the results of Table 5.10, a plot of x_m versus $1/kT_d$ gives the trapping parameters according to equation (5.43). The values obtained are $E = 0.993 \pm 0.003$ and $s = (8.11 \pm 0.91) \, 10^{11} \, s^{-1}$, which are very close to the original parameters of $E = 1$ eV and $s = 10^{12} \, s^{-1}$.

(c) Using equation (5.37), the phosphorescence intensity (I) versus time (t) are evaluated for different values of the kinetic order b. The respective TL-like presentation curves can be obtained either by the transformation of $x = \ln(t)$ and $y = It$ or directly from equation (5.40). Examples of exponential and TL-like phosphorescence decay curves for values of b between $b = 1$ and $b = 2$ are shown in Figure 5.13. The exponential decay is plotted in semilog scale. Once again

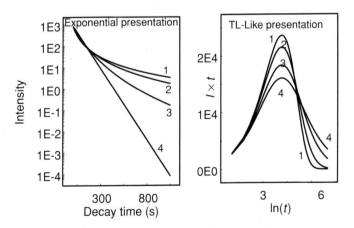

FIGURE 5.13. Exponential and TL-like presentation of general-order phosphorescence decay curves (1) $b = 1$, (2) $b = 1.2$, (3) $b = 1.6$ and (4) $b = 2$.

Table 5.11. The values of the geometrical shape factor μ_g versus kinetic order b evaluated from the simulated TL-like presentation curves

b	1	1.1	1.2	1.3	1.4	1.5	1.6	1.7	1.8	1.9	2
μ_g	0.4099	0.4235	0.4359	0.4473	0.4578	0.4672	0.4767	0.4851	0.4932	0.5008	0.5080

the exponential presentation appears featureless when compared with the TL-like representation of the data.

By using an analysis similar to the one employed for TL glow curves, the peak-shaped TL-like presentation curves can be characterized by the peak maximum position x_m, as well as by the half maximum intensities x_{m1} and x_{m2}, respectively. One can also define the width parameters $\omega = x_{m2} - x_{m1}$, $\delta = x_{m2} - x_m$, $\tau = x_m - x_{m1}$ and finally the symmetry factor $\mu_g = \delta/\omega$.

The values of these parameters are easily evaluated from the simulated TL-like presentation curves. The values of the geometrical shape factor μ_g versus kinetic order b are listed in Table 5.11 and shown in Figure 5.14. According to these results, the relation between the symmetry factor and the kinetics order of the TL-like presentation phosphorescence decay curves is almost identical with the respective behavior of these parameters in the usual TL glow peak (Figure 1.15).

(d) If thermal quenching is present, the exponential decay curve for first-order kinetics is given by the expression.

$$I_{QU} = \eta(T_d)I_0 \exp\left(-\frac{t}{\tau(T_d)}\right), \qquad (5.46)$$

where T_d is the decay temperature and $\eta(T_d)$ is the quenching efficiency at the decay temperature T_d.

In order to evaluate the trapping parameters E and s_{eff}, one has to obtain equation (5.46) at various decay temperatures T_d, and to evaluate $\tau(T_d)$ from the slope of the graph on a semilog scale. However, since $\eta(T_d)$ has unique values at each temperature T_d, its influence is restricted in changing the intercept from a value of

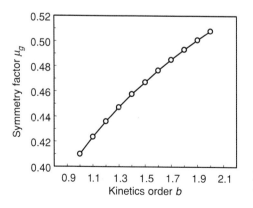

Figure 5.14. Symmetry factor μ_g versus kinetic order b.

TABLE 5.12. The corrected peak maximum T_{Mc} and the temperature lag ΔT

$\beta(K s^{-1})$	T_{M1} (K)	T_{Mc} (K)	ΔT(K)	$E(T_{M1})$ (eV)	$E(T_{Mc})$ (eV)	T_{Me} (K)	ΔT (K)
1	481.5	481.5	0	0	0	481.5	0
2	489.0	489.0	0	1.96	1.96	488.4	0.64
3	494.8	493.4	1.4	1.78	1.98	492.4	2.43
4	499.2	496.5	2.7	1.71	1.99	495.2	3.98
8	510.2	504.0	6.2	1.62	2.02	502.1	8.14
12	518.2	508.4	9.8	1.54	2.03	506.1	12.11
18	529.8	512.8	17.0	1.40	2.05	510.1	19.69
25	548.2	516.33	31.9	1.19	2.07	513.3	34.84

$\ln(I_0)$ in the case of no quenching to a value of $\ln(\eta(T_d)I_0)$ in case of quenching, whereas the value of the slope remains the same.

This means that the evaluation of the trapping parameters E and s_{eff} is not influenced by the presence of thermal quenching when using phosphorescence data. Therefore, the phosphorescence decay method is the only available method of trapping parameter evaluation, which is not influenced by thermal quenching.

Exercise 5.10: Temperature Lag Corrections

During experimental TL work, the temperature measured is usually that of the heating element as measured by a thermocouple attached to it. In order to perform a TL glow curve analysis and extract meaningful parameters from the data, it is imperative to know the true temperature of the sample. Experimentally, it is found that there is a temperature lag between the thermocouple measurement and the actual temperature of the sample. The purpose of this exercise is to show how to correct experimental TL data for this *temperature lag effect*.

You are given the experimental peak maximum T_{Mg} of peak 5 of LiF:Mg, Ti as a function of the heating rate used to measure the glow curve. The heating rate is in column 1 of Table 5.12 and the peak maxima are shown in column 2. In graphical form, the T_{Mg} values are shown in curve (a) of Figure 5.15.

(a) Find the corrected peak maxima T_{Mc} by applying a correction due to the temperature lag effect. Evaluate the temperature lag ΔT at the peak maximum position.

(b) Apply the variable heating rate method using the uncorrected peak maxima T_{Mg} and also by using the corrected T_{Mc} values. Comment on the observed differences.

(c) Find the activation energy E for this TL glow peak by applying the two heating rate method of analysis, and study the influence of the temperature lag effect on the value of E.

(d) If the value of E is known to be $E = 2.05$ eV, find again the temperature lag under this assumption, and compare with the values of ΔT obtained in (a).

(e) Discuss some of the experimental factors which can influence the temperature lag between the sample and the heating element.

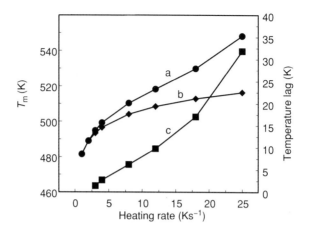

FIGURE 5.15. Left y-axis: (a) Experimental values T_{Mg} (b) Corrected values T_{Mc}, as functions of the heating rate. Right y-axis: Temperature lag ΔT as a function of the heating rate.

Solution

(a) During TL readout with readers using contact heating, the temperature of the sample differs from the temperature of the heating element. This difference is called temperature lag. A simple approximate method to find the temperature lag based on TL measurements only is described by the following equation [14]:

$$T_{Mj} = T_{Mi} - c \cdot \ln\left(\frac{\beta_i}{\beta_j}\right), \tag{5.47}$$

where T_{Mj} and T_{Mi} are the maximum temperatures of the glow peak with rate of heating β_j and β_i, respectively, and c is a constant. The experimental data were obtained using ^{6}LiF:Mg,Ti chips of dimensions $3 \times 3 \times 0.9$ mm. The measurements were recorded with a Harshaw model 3500 manual TLD reader with continuous nitrogen flow. The test dose was 12 mGy from a beta-ray source.

The steps required to evaluate the temperature lag are as follows:

Step 1: The constant c is evaluated using two very low heating rates, where the temperature lag is assumed to be zero. In practice, however, solid chip samples cannot completely avoid the temperature lag effect, even at these low heating rates. Using the T_{Mg} temperatures corresponding to the heating rates 1 and 2 K s^{-1} from Table 5.12, the constant c can be found by using equation (5.47):

$$c = \frac{T_{\text{M2}} - T_{\text{M1}}}{\ln(2)} = 10.8202. \tag{5.48}$$

Step 2: Using equation (5.47) and the evaluated constant c, the corrected peak maximum T_{Mc} at every heating rate β is estimated and listed in column 3 of Table 5.12. These corrected peak temperatures are also shown as curve (b) in Figure 5.15.

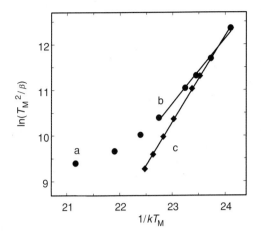

FIGURE 5.16. (a) $\ln(T_{\text{Mg}}^2/\beta)$ versus $1/kT_{\text{Mg}}$ plot with the experimental data. (b) The sample plot using the data for heating rates $1-8\,\text{K s}^{-1}$. (c) The same plot using the temperature lag corrected values T_{Mc}.

Step 3: The temperature lag ΔT at the position of the peak maximum is $T_{\text{Mg}} - T_{\text{Mc}}$ and is given in column 4 of Table 5.12, and in graphical form as curve (c) in Figure 5.15.

(b) The variable heating rate method (see Exercise 2.2 in Chapter 2) consists of a plot of $\ln(T_M^2/\beta)$ versus $1/kT_M$. This plot should be theoretically a straight line with slope E and intercept $A = \ln(E/s\,k)$, from which the pre-exponential factor s can be calculated using $s = (E/k)\exp(-A)$.

The plots of $\ln(T_M^2/\beta)$ versus $1/kT_M$ are shown in Figure 5.16 by using both the uncorrected experimental data, as well as using the corrected peak temperatures T_{Mc}. Figure 5.16 shows that the experimental values of T_{Mg} (solid circles) do not lead to a straight line due to the temperature lag effect.

By using the uncorrected experimental data for heating rates $1-8\,\text{K s}^{-1}$, a very good line is obtained ($R = 0.994$), from which it is found that $E = 1.429 \pm 0.066$ eV and $A = -22.1694 \pm 1.559$. The calculated value of $s = 7.043 \times 10^{13}\,\text{s}^{-1}$.

By using the corrected peak temperatures T_{Mc}, a very good straight line is obtained ($R = 0.9996$) for all the data points. From the best fitted line through the corrected data, one obtains $E = 1.898 \pm 0.016$ eV, $A = -33.3549 \pm 0.036$, and the calculated $s = 6.742 \times 10^{18}\,\text{s}^{-1}$.

The conclusion is that when using the uncorrected T_{Mg} data, the activation energy E is seriously underestimated. On the other hand, by using the corrected T_{Mc} data the values of activation energy approaches sufficiently the known values of E, which is between 1.9 and 2.1 eV.

(c) In the two-heating-rate method the activation energy is evaluated by the equation [15]:

$$E = \frac{kT_{M1}T_{M2}}{T_{M1} - T_{M2}} \ln\left(\frac{\beta_1}{\beta_2} \cdot \frac{T_{M1}^2}{T_{M2}^2}\right). \tag{5.49}$$

Assuming that T_{M1} is the temperature corresponding to the lowest available heating rate of $1\,\text{K s}^{-1}$, and by substituting for T_{M2} each of the higher heating rates shown in Table 5.12 successively, the resulting values of E are listed in columns 5 and 6 of Table 5.12, for the uncorrected and corrected peak maxima, respectively. The results listed in column 5 of Table 5.12 show clearly the extreme influence of the temperature lag on the activation energy values obtained by the two heating rate method of analysis.

On the other hand, the results of column 6 show the effectiveness of the temperature lag correction.

(d) If the activation energy E of glow peak is known, then the calculation of the temperature lag effect can be achieved more accurately as follows. In this case, the constant c can be evaluated by the equation [14]:

$$c = T_{Mj} T_{Mi} \frac{k}{E}, \qquad (5.50)$$

which involves the known activation energy of the peak.

By using the two lowest available heating rates ($i = 1$ and $j = 2$) from column 2 of Table 5.12, one finds $c = 9.8971$. Using equation (5.47), the new peak maxima T_{Me} are evaluated and listed in column 7 of Table 5.12.

The new values of the temperature lag are equal to $T_{Mg} - T_{Me}$ and are listed in column 8. It is interesting to observe that at the heating rate of 2 K s^{-1}, a $\Delta T = 0.64$ K is detected, whereas initially this temperature lag was assumed to be zero.

By applying the variable heating rate method to the data of column 7, it is found that $E = 2.07 \pm 0.016$ eV, i.e. 1% difference with the used value of 2.05 eV. The intercept is $A = -37.5145 \pm 0.367$ from which the value of $s = 4.55 \times 10^{20}$ s^{-1}.

(e) The critical point of the temperature lag correction method is to avoid the temperature lag at the lowest available heating rates. This can be achieved by using silicon oil of very high thermal conductivity between the heating element and the chip. Even better, one can use loose powder instead of chips, with the silicon oil applied on the heating element. The extrapolation of loose powder results to the chip is permitted because the correction method described here involves the glow peak positions only.

Some of the additional factors that can influence the temperature lag effect are as follows:

1. Temperature gradients across the heating element.
2. Temperature gradients across the sample itself.
3. Nonideal thermal contact between the heating element and the sample.
4. Effects of the inert atmosphere in the TL chamber.

Exercise 5.11: Study of the Integrals Appearing in the Expressions for First- and General-Order TL Kinetics

The purpose of this exercise is to study the various terms that appear in the equations describing first-order and general-order TL glow peaks. By performing a numerical analysis of these terms for a wide range of the activation energies E and the frequency factors s, a formula is derived for the activation energy E as a function of the FWHM ω and the temperature T_M of maximum TL intensity. This formula is compared with the well-known equation (1.49).

The analytical expression of a first-order single glow peak with the series approximation is given by equation (3.26):

$$I(T) = n_0 \, s \exp\left(-\frac{E}{kT}\right) \exp\left\{-\frac{skT^2}{\beta E} \exp\left(-\frac{E}{kT}\right)\left(1 - \frac{2kT}{E}\right)\right\}. \quad (5.51)$$

The corresponding equation for general-order kinetics is (3.33):

$$I(T) = n_0 \, s \exp\left(-\frac{E}{kT}\right)\left[1 + (b-1)\frac{skT^2}{\beta E} \exp\left(-\frac{E}{kT}\right)\left(1 - \frac{2kT}{E}\right)\right]^{-\frac{b}{b-1}}. \quad (5.52)$$

Assuming for simplicity that $n_0 = 1$, these analytical expressions consist of two parts $F1(T)$ and $F2(T)$ such that:

$$F1(T) = s \exp\left(-\frac{E}{kT}\right) \quad (5.53)$$

First-order $$F2(T) = \exp\left\{-\frac{skT^2}{\beta E} \exp\left(-\frac{E}{kT}\right) \cdot \left(1 - \frac{2kT}{E}\right)\right\} \quad (5.54)$$

General-order $$F2(T) = \left\{1 + \frac{(b-1)skT^2}{\beta E} \exp\left(-\frac{E}{kT}\right)\left(1 - \frac{2kT}{E}\right)\right\}^{-\frac{b}{b-1}}. \quad (5.55)$$

The first function $F1(T)$ is the well-known Boltzmann function and is an increasing function of temperature, while the function $F2(T)$ is a decreasing function of temperature. The product of the two functions yields the peak-shaped graph $I(T)$ for the TL intensity.

(a) Find an expression for the peak maximum intensity I_M for both first- and general-order kinetics in terms of E and T_M.
(b) Find an expression for $F1(T)$ and for $F2(T)$ in terms of E and T_M.
(c) Evaluate synthetic glow peaks of any kinetic order using equations (5.51)–(5.52) and (5.51)–(5.53) and obtain the values of I_M, T_M, ω, and μ_g for arbitrary pairs of E and s values. Investigate the degree of variation of the functions $F1(T)$ and $F2(T)$ for arbitrary E and s pairs. Discuss the properties of these functions.
(d) Find expressions for the activation energy E based on the peak shape of the glow curves, in terms of the width ω of the glow curve.
(e) Discuss the findings and find a general peak shape expression for the activation energy.
(f) Compare the peak shape expression derived in this exercise with the respective equation (1.48) derived by Chen [16].

Solution

(a) The maximum intensity I_M is obtained by replacing $T = T_M$ in equation (3.26):

$$I_M = s \exp\left(-\frac{E}{kT_M}\right) \exp\left\{-\frac{skT_M^2}{\beta \cdot E} \exp\left(-\frac{E}{kT_M}\right)\left(1 - \frac{2kT_M}{E}\right)\right\} \quad (5.56)$$

and by replacing $T = T_M$ in equation (3.33) for general-order kinetics:

$$I_M = s \exp\left(-\frac{E}{kT_M}\right)\left\{1 + \frac{(b-1)skT_M^2}{\beta \cdot E}\exp\left(-\frac{E}{kT_M}\right)\left(1 - \frac{2kT_M}{E}\right)\right\}^{-\frac{b}{b-1}}. \quad (5.57)$$

The conditions for the maximum are given by equations (1.8) and (1.10):

First-order kinetics $\quad \dfrac{\beta E}{kT_M^2} = s \exp\left(-\dfrac{E}{kT_M}\right) \quad$ (5.58)

General-order kinetics $\quad \dfrac{\beta E}{kT_M^2} = \left(1 + (b-1)\dfrac{2kT_M}{E}\right) s \exp\left(-\dfrac{E}{kT_M}\right). \quad$ (5.59)

By substituting equations (1.8) and (1.10) into equations (5.56) and (5.57) normalized over the heating rate β leads to the following expressions for I_M, respectively [14–15]:

First-order kinetics $\quad I_M = \dfrac{E}{kT_M^2} \cdot \dfrac{1}{e} \cdot \exp\left(\dfrac{2kT_M}{E}\right) \quad$ (5.60)

General-order kinetics $\quad I_M = \dfrac{E}{kT_M^2} \cdot \left\{\dfrac{b}{1 + (b-1) \cdot \dfrac{2kT_M}{E}}\right\}^{-\frac{b}{b-1}}. \quad$ (5.61)

(b) Using equations (5.58) and (5.59), the function $F2(T)$ from equations (5.54) and (5.55) at the maximum position T_M becomes, respectively:

First-order kinetics $\quad F2(T_M) = \dfrac{1}{e} \exp\left(\dfrac{2kT_M}{E}\right) \quad$ (5.62)

General-order kinetics $\quad F2(T_M) = \left[\dfrac{b}{1 + (b-1) \cdot \dfrac{2kT_M}{E}}\right]^{-\frac{b}{b-1}}. \quad$ (5.63)

(c) Using a spreadsheet program, we can easily evaluate synthetic glow peaks of any kinetic order, using equations (5.51) and (5.52), in order to obtain T_M, I_M, ω, and the symmetry factor μ_g. By considering any arbitrary pair E and s in the range $E = 0.5$–2 eV and $s = 10^9$–10^{20} s^{-1}, we will find that the functions $F_1(T_M)$

TABLE 5.13. Values of the functions $F1(T)$, $F2(T)$, and the pseudo-constants C_b, C_d, and C_f

b	μ_g	$F1(T), F2(T) = C_b$	$I\omega = C_d$	$C_f = C_d/C_b$
1	0.41416 ± 0.0025	0.38937 ± 0.0042	0.91762 ± 0.0038	2.3567 ± 0.0272
1.1	0.43011 ± 0.0025	0.37302 ± 0.0044	0.91585 ± 0.0036	2.4552 ± 0.0305
1.2	0.44279 ± 0.0026	0.35837 ± 0.0045	0.91343 ± 0.0033	2.5488 ± 0.0333
1.3	0.45438 ± 0.0026	0.34432 ± 0.0047	0.91054 ± 0.0030	2.6445 ± 0.0370
1.4	0.46506 ± 0.0027	0.33317 ± 0.0049	0.90729 ± 0.0028	2.7232 ± 0.0409
1.5	0.47496 ± 0.0027	0.32222 ± 0.0050	0.90374 ± 0.0025	2.8047 ± 0.0442
1.6	0.48418 ± 0.0027	0.31222 ± 0.0052	0.90000 ± 0.0022	2.8826 ± 0.0485
1.7	0.49280 ± 0.0027	0.30298 ± 0.0053	0.89613 ± 0.0019	2.9577 ± 0.0521
1.8	0.50087 ± 0.0027	0.29439 ± 0.0054	0.89213 ± 0.0016	3.0304 ± 0.0559
1.9	0.50852 ± 0.0028	0.28652 ± 0.0055	0.88808 ± 0.0014	3.0995 ± 0.0597
2	0.51575 ± 0.0027	0.27916 ± 0.0056	0.88397 ± 0.0011	3.1615 ± 0.0636

and $F_2(T_M)$ vary extremely slowly. For a given order of kinetics, these functions can be considered pseudo-constants, let us say C_b. For $b = 1$, $C_b = F_1(T_M)$ and for $b > 1$, $C_b = F_2(T_M)$.

The reader can evaluate the values of the pseudo-constant C_b for various kinetic orders and verify that the value of C_b will be within the ranges given in Table 5.13.

(d) Using the results from (c), equations (5.60) and (5.61) can be solved for the activation energy E for any kinetics order to yield:

$$E = \frac{I_M}{C_b} kT_M^2. \tag{5.64}$$

Using the parameter $\omega = T_2 - T_1$, equation (5.64) becomes

$$E = \frac{I_M \cdot \omega}{C_b} \cdot \frac{kT_M^2}{\omega}. \tag{5.65}$$

Observing equation (5.65) one can argue that in order to have a peak shape formula similar to the very well-known Chen expressions, the following requirement has to be fulfilled:

$$I_M \cdot \omega = \text{Constant} = C_d. \tag{5.66}$$

Equation (5.66) does indeed hold for any pair of (E, s) values and for any kinetic order b. It is nothing else than the well-known triangle assumption made in the past in order to obtain the peak shape methods. The values given in the literature are 0.92 and 0.88 for first- and second-order kinetics, respectively [15]. The reader can evaluate the values of C_d from the derived synthetic glow peaks and verify that C_d will be within the values given in Table 5.13.

Therefore, the new form of equation (5.65) is

$$E = \frac{C_d}{C_b} \cdot \frac{kT_M^2}{\omega}. \tag{5.67}$$

Since the ratio of two pseudo-constants will be a pseudo-constant as well, equation (5.67) is rewritten as

$$E = C_f \cdot \frac{kT_M^2}{\omega}. \qquad (5.68)$$

(e) Discussion of the findings

The symmetry factor μ_g is a quantity that can be obtained experimentally from the measured glow curves. Therefore, it is of interest to study the behavior of the pseudo-constants as a function of μ_g. The pseudo-constant C_b is a linear function of the symmetry factor μ_g as shown in Figure 5.17. The linear function of Figure 5.17 is

$$C_b = (0.8418 \pm 0.032) - (1.0927 \pm 0.0068)\mu_g. \qquad (5.69)$$

The behavior of the pseudo-constant C_d is shown in Figure 5.18, where one observes that the pseudo-constant C_d varies by only 4%, whereas C_b varies by 30% from b $= 1$ to b $= 2$. This means that the behavior of the ratio $C_f = C_d/C_b$ should be governed mainly by the behavior of C_b. Therefore, C_f should also be a linear function of μ_g, as shown in Figure 5.19. The linear relationship obtained from Figure 5.19 is

$$C_f = (-1.0593 \pm 0.0677) + (8.1641 \pm 0.1409)\mu_g. \qquad (5.70)$$

By combining equations (5.68) and (5.70), a new expression for the activation energy which holds for any kinetics order b is deduced, i.e

$$E = (8.1641\,\mu_g - 1.0593)\frac{kT_M^2}{\omega}. \qquad (5.71)$$

(f) The respective Chen equation (1.49) is

$$E_c = (2.52 + 10.2(\mu_g - 0.42)) \cdot \frac{kT_M^2}{\omega} - 2kT_M. \qquad (5.72)$$

The difference $\Delta E = E_c - E$ between (5.71) and (5.72) after some algebra is found to be

$$\Delta E = (2.0359 - 0.7047)\frac{kT_M^2}{\omega} - 2kT_M \qquad (5.73)$$

or

$$\Delta E = kT_M\left[(2.0359\mu_g - 0.7047)\frac{T_M}{\omega} - 2\right]. \qquad (5.74)$$

From equation (5.74), it seems that the difference becomes zero when

$$\frac{T_M}{\omega} = \frac{1}{1.0179\mu_g - 0.352}. \qquad (5.75)$$

Equation (5.75) gives $T_M/\omega = 13$ for $\mu_g = 0.42$ and 5.7 for $\mu_g = 0.52$ and intermediate values for the other μ_gs. Therefore, the agreement of the present peak shape expression with that of Chen's depends on the specific glow peak.

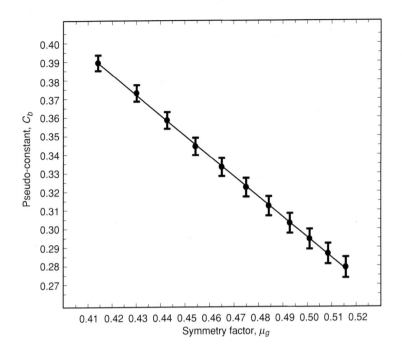

FIGURE 5.17. Pseudo-constant C_b versus symmetry factor.

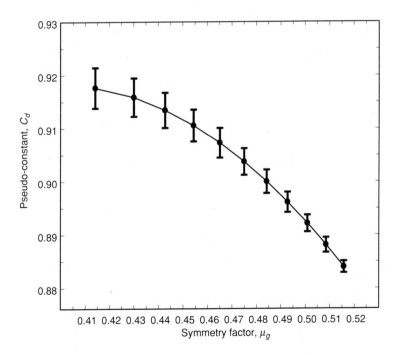

FIGURE 5.18. Pseudo-constant C_d versus symmetry factor.

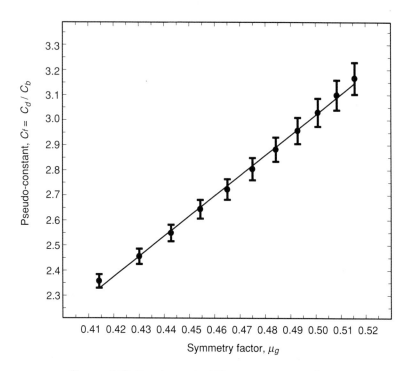

FIGURE 5.19. Pseudo-constant C_f versus symmetry factor.

From Equation (5.74), it can be found that the highest deviation of the present formula relative to that of Chen for first-order kinetics with $\mu_g = 0.42$, is $(\Delta E / E) = 0.15/2.52 = 5.9\%$ and for second-order kinetics with $\mu_g = 0.52$, is $(\Delta E / E) = 0.353/3.54 = 9.9\%$.

For a more detailed study of the subject in this exercise and the development of acceptance criteria for E and s values, see the work by Kitis [17].

References

[1] B. Burkhardt and E. Piesh, *Nucl. Instr. Meth.* **175** (1980) 159.
[2] P. Zarand and I. Polgar, *Nucl. Instr. Meth.* **205** (1983) 525.
[3] P. Zarand and I. Polgar, *Nucl. Instr. Meth.* **222** (1984) 567.
[4] C. Furetta, 2003. *Handbook of Thermoluminescence*. World Scientific: Singapore.
[5] P. Plato and J. Miklos, *Health Phys.* **49** (5) (1985) 873.
[6] C. Furetta, 2003. *Handbook of Thermoluminescence* . World Scientific: Singapore, pp. 138.
[7] M.S. Akselrod, N. Agersnap Larsen, V. Whitley, and S.W.S. McKeever, *Journal of Applied Physics* **84** (1998) 3364.
[8] R. Chen, *J. Electrochem. Soc.* **116** (1969) 116.
[9] S.A. Petrov and I.K. Bailiff, *J. of Lumin.* **65** (1996) 289.
[10] G. Kitis, *Phys. Stat. Sol. (a)* **191** (2002) 621.

[11] J. T. Randal and M.H.F. Wilkins, *Proc. R. Soc.* (London) **A 184** (1945) 390.

[12] Y. Kirsh and R. Chen, *Nucl. Tracks Radiat. Meas.* **18** (1991) 37.

[13] R. Chen and S.W.S. McKeever, 1997. *Theory of Thermoluminescence and Related Phenomena.* World Scientific, p. 97–101.

[14] G. Kitis and J.W.N. Tuyn, *J. Phys. D: Appl. Phys.* **26** (1998) 2065.

[15] R. Chen and S.A.A. Winer, *J. Appl. Phys.* **41** (1970) 5227 (and references herein).

[16] R. Chen, J. Appl. Phys. **40** (1969) 570–585.

[17] G. Kitis, *J. Radioanal. Nucl. Chem.* **247** (2001) 697–703.

Annotated Bibliography

The textbooks, monographs, and papers included in this annotated bibliography were chosen by their relevance to the topics covered in this book. The list is not meant to be comprehensive but rather to provide a starting point for researchers and graduate students in the field. They are listed in *reverse chronological order* and comprise the following categories:

1. Thermoluminescence books
2. Papers dealing with numerical methods used in TL data analysis
3. Papers describing kinetic models in TL
4. TL versus dose dependence papers
5. Review papers in TL
6. Papers on curve fitting and deconvolution functions
7. Papers on thermal quenching and Temperature lag effects

1. Thermoluminescence Books

Handbook of Thermoluminescence

C. Furetta
World Scientific Publishing Co., Singapore, 2003.
This book on TL provides experts, teachers, students, and technicians practical support for research, study, and routine work. Special effort has been made to include the TL terminology commonly used in the literature. The topics are given in alphabetical order to facilitate searching for topics. The topics covered are various TL models, methods for determining the kinetic parameters, procedures for characterizing a thermoluminescent dosimetric system, and others. The analytical treatments of TL models are fully developed.

Radiation Dosimetry-Instrumentation and Methods, 2nd Edition

Gad Shani
CRC Press, 2000.
This is an extensive reference book for medical applications of dosimetry, and is an assembly of various developments in the field. After two chapters describing theoretical aspects of dosimetry and radiation interactions, an extensive chapter is dedicated to the properties and uses of ionization chambers. The fourth chapter in the book covers broad aspects of thermoluminescence dosimetry (TLD), with emphasis on commonly used dosimetric materials like LiF:Mg,Ti. The rest of the book covers several other techniques used in dosimetry, like the radiographic film, 3-D dosimetry and neutron dosimetry.

Operational Thermoluminescence Dosimetry

C. Furetta and P.S. Weng
World Scientific, Singapore, 1998.
This small book is mainly derived from a course on thermoluminescence given by one of the Authors (C. Furetta) at the National Tsing Hua University, Taiwan, in 1982/83. The main features of the book are the mathematical treatment of the various theories of thermoluminescence, as well as of the experimental methods used to evaluate the TL parameters. Another important part of the book covers the procedures for the set up of a thermoluminescent dosimetric system, as well as the factors used in the dose determination from the thermoluminescent emission. Others parts of the book include (i) the precision and the accuracy in TL measurements, (ii) practical procedures for environmental, personal, and clinical dosimetry, and (iii) lower detection limit. The book is of a practical nature and can be very useful for students and technicians in the field of thermoluminescence, as well as form the basis for a course in solid-state dosimetry.

An Introduction to Optical Dating: The Dating of Quaternary Sediments by the Use of Photon-Stimulated Luminescence

M.J. Aitken
Oxford University Press, 1998.
The book discusses optical dating, a rapidly developing technique which is used primarily in the dating of sediments deposited in the last 500,000 or more years. The book is divided into three parts consisting of the main text, the technical notes, and the appendices. The book introduces the method with characteristic applications, and discusses the limitations of the optical dating technique.

Theory of Thermoluminescence and Related Phenomena

R. Chen and S.W.S. McKeever
World Scientific Publishing Co., Singapore, 1997.
This is the authoritative and most informative book on thermoluminescence written by two of the world's experts on the topic. After an introductory chapter on thermally stimulated processes, three chapters are devoted to TL, analysis of TL glow curves and the nonlinear dose dependence of TL. Fifth chapter and a chapter on TL applications, covers optical phenomena, followed by mathematical considerations on solving differential equations and peak fitting. Two chapters are on related phenomena and simultaneous measurements of TL and other thermally stimulated process. Final chapter covers miscellaneous effects related to TL.

Thermoluminescence Dosimetry Materials: Properties and Uses

S.W.S. McKeever, M. Moscovitch, and P.D. Townsend
Nuclear Technology Publishing, UK, 1995.
This book provides a review of the TL properties and dosimetric properties of the most common TLD materials. The emphasis in the book is placed on making links between the solid-state defects in these materials and their dosimetry properties. After an introductory chapter giving a broad description of the TL process and TL dosimetry, three extensive separate chapters are dedicated to Fluorides, Oxides, and Sulphates/Borates.

Thermoluminescent Materials

D.R. Vij, Editor
PTR Prentice Hall, New Jersey, 1993.
This book provides a very broad review of materials exhibiting TL of practical use. The types of materials covered range from natural materials like minerals and quartz, to organic materials like polymers and finally to a wide range of inorganic materials. Methods of preparation for several materials and list of their characteristics and uses are provided. Each chapter also contains a list of applications for the materials covered in the chapter.

Thermoluminescence in Solids and Its Applications

K. Mahesh, P.S. Weng, and C. Furetta
Nuclear Technology Publishing, UK, 1989.
The aim of the book is to offer a comprehensive study of the features of thermoluminescence from both a theoretical and a practical point of view. After a chapter dedicated to the historical background, the second chapter covers the general properties of luminescence phenomena, and the principles and methods of thermoluminescence. Two chapters are dedicated to TL materials and instrumentation, and

a short chapter introduces the models and the theories of thermoluminescence. A very interesting part of the book is dedicated to the TL-related phenomena, i.e. ES, TSEE, TSC, TSD. The last two chapters include applications and developments in TL. One of the appendices covers the phosphor terminology. The book can be very useful for teachers and students in solid-state physics, nuclear science, and radiation dosimetry, although the more recent TL models are not included due to its year of publication.

Thermoluminescence Dating

M.J. Aitken
Academic Press, 1985.
This book is a comprehensive introduction to TL dating covering pottery dating, natural and artificial irradiation of samples, special dating methods, TL methods that can be used for other types of materials, and sediment dating. Even though some of the topics may be dated, it is still a valuable and clearly written introduction to TL dating.

Thermoluminescence of Solids

S.W.S. McKeever
Cambridge University Press, UK, 1985.
This is the first book that presented thermoluminescence from a solid-state physics point of view and for this reason it is very important. The aim of the book is to unify the various aspects of thermoluminescence, i.e. dating, dosimetry, kinetics studies, etc. The starting point of the book is the description of thermolumines-cence within the context of luminescence phenomena. A theoretical background follows which includes the elementary concepts and the various models used in TL. An important chapter takes into consideration the relationship between crystal defects and thermoluminescence. More or less half of the book covers thermolu-minescent materials and their application in personal, environmental, and medical dosimetry. Other subjects of the book include dating, geological applications of TL and instrumentation.

Thermoluminescence and Thermoluminescent Dosimetry, Vols. I, II, and III

Y.S. Horowitz
CRC Press, USA, 1984.
This extensive three-volume book covers most theoretical and applied aspects of TL. The first volume covers general aspects of TL, TL kinetic models, and lists of important TL dosimetric materials and their properties. The second volume covers TL versus dose response and TL models for superlinearity and sensitization and

the uses of TLDs for various radiation fields, and track structure theory. The final
volume covers instrumentation and applications of TL.

Analysis of Thermally Stimulated Processes

R. Chen and Y. Kirsh
Pergamon Press, USA, 1981.
This book is one of the most important publications in the field of the thermally
stimulated processes. The Authors cover the following thermally stimulated processes: Thermoluminescence (TL), thermally stimulated conductivity (TSC), thermally stimulated electron emission (TSEE), thermally stimulated depolarization
(TSD), thermogravity (TG), derivative thermogravity (DTG), and differential thermal analysis (DTA). The book is presented as an interdisciplinary text. The theories
concerning first-, second-, and general-order kinetics in TL are fully developed and
explained. A full chapter is dedicated to the various methods used in the evaluation
of the kinetics parameters from the thermally stimulated curves. Each chapter has
a very large list of references.

Thermoluminescence Dosimetry

A.F. McKinley
Adam Hilger Ltd., Bristol, UK, 1981.
Persons who had the opportunity to meet Alastair McKinley are familiar with his
very clear method of explaining scientific subjects. This small book provides a
clear introduction to the use of TLDs in ionizing radiation measurements, with
particular emphasis in clinical dosimetry. The theory of thermoluminescence is
described briefly and only first-order kinetics is taken into consideration. The
most important parts of the book are:

(i) The use of TLDs for specific applications such as clinical, personal, environmental, charged particle, neutron, and mixed-field dosimetry;
(ii) Experimental problems regarding the annealing procedures, the storage and
handling of TLDs and their irradiation.

Although the book of McKinley is more than 20 years old, it contains many
practical suggestions that are still useful.

Thermoluminescence: Its Understanding and Applications

K.S.V. Nambi
Published by Instituto de Energia Atomica, Cidade Universitaria Armando de
Salles Oliveira, São Paulo, Brasil, 1977.
The book by Nambi, of about 100 pages in A4 size, can be considered the first
effort to present together the known aspects of thermoluminescence at that time.
It is surprising how many aspects of TL are included in this book, which would
normally be found dispersed in hundreds of scientific publications. One of the

most interesting parts of the book is that it considers the various factors affecting TL, effects which very often are not taken into consideration. The book also contains interesting information concerning the use of TL in the forensic sciences.

2. Papers Dealing with Numerical Methods Used in TL Data Analysis

Limitation of Peak Fitting and Peak Shape Methods for Determination of Activation Energy of Thermoluminescence Glow Peaks

C.M. Sunta, W.E.F. Ayta, T.M. Piters, and S. Watanabe, *Radiat. Meas.* **30** (1999) 197.
This paper investigates the validity of peak shape methods of TL analysis and of peak-fitting techniques under two conditions (a) when the retrapping probability is much higher than the recombination probability and (b) when the traps are filled near the saturation level. Examples of calculations are given within the OTOR and IMTS models of thermoluminescence, and it is recommended that the peak shape and peak-fitting methods can be applied only at low doses, far from TL saturation conditions.

Anomalies in the Determination of the Activation Energy of Thermoluminescence Glow Peaks by General-Order Fitting

C.M. Sunta, W.E.F. Ayta, T.M. Piters, R.N. Kulkarni, and S. Watanabe, *J. Phys. D: Appl. Phys.* **32** (1999) 1271.
Several TL glow curves are calculated within the OTOR, NMTS, and IMTS models and under the quasi-equilibrium (QE) conditions. These glow curves are analyzed by fitting them using the empirical general-order (GO) kinetics model, in order to find the parameters b and E. It is shown that the fitted value of E and the quality of fit (figure of merit, FOM), depart from the expected values as b deviates from 1 to 2. These results are applied to the interpretation of the E values obtained experimentally for peak 5 of LiF (TLD − 100).

Analysis of the Blue Phosphorescence of X-Irradiated Albite Using a TL-Like Presentation

Y. Kirsh and R. Chen, *Nucl. Tracks Radiat. Meas.* **18** (1991) 37.
This paper describes a procedure by which a featureless exponential decay curve is transformed into a peak-shaped curve, which resembles the corresponding thermoluminescence curve. The paper treats first- and general-order decay curves and derives analytical expressions for the corresponding peak-shaped curves. These expressions can be used for a direct fit of the experimental decay curves in

$I(T) \times t$ versus $\ln(t)$ scale, from which trapping parameters like activation energy, frequency factor, and kinetic order can be obtained. Furthermore, this kind of treatment of the experimental isothermal decay curves can be successfully extrapolated to the optical simulated luminescence decay curves.

Determination of Thermoluminescence Parameters from Glow Curves: II in CaSo4:Dy

J. Azorin and A. Gutierrez
Nucl. Tracks **11** (1986) 167.
This paper is a good example of a comprehensive analysis of TL glow curves which follow second-order kinetics. A wide variety of methods is used to analyze the TL glow curves and the results of the different methods are in good agreement. Isothermal methods, heating rate methods, initial rise, peak shape, and thermal cleaning methods are used. The E and s values for the curves are calculated and the results of different methods are compared with each other.

A Theoretical Study on the Relative Standard Deviation of TLD Systems

P. Zarand and I. Polgar, *Nucl. Instr. Methods* **205** (1983) 525.

On the Relative Standard Deviation of TLD Systems

P. Zarand and I. Polgar, *Nucl. Instr. Methods* **222** (1984) 567.
In these two papers the authors propose a model to describe the relative standard deviation of the TL readings obtained after a given dose. The different behaviors of TLD systems in the low-dose range are also discussed. In the second paper the theoretical model is submitted to experimental verification.

Analysis of Thermoluminescence Data Dominated by Second-Order Kinetics

R. Chen, D.J. Huntley, and G.W. Berger, *phys. stat. sol. (a)* **79** (1983) 251.
This paper describes the TL response for peaks following second-order kinetics. The plateau test is applied to calculated second-order glow peaks, and also for the case of a distribution of second-order peaks. The paper also contains a very useful set of criteria indicating second-order kinetics in TL experiments.

Reproducibility of TLD Systems. A Comprehensive Analysis of Experimental Results

B. Burkhardt and E. Piesh, *Nucl. Instr. Methods* **175** (1980) 159.
The paper presents a study of the statistical errors involved in low-dose measurements of TLD systems. The analysis takes into account the dark current, the

zero dose reading, and the irradiation and annealing procedures. A two-parameter expression of the standard deviation versus exposure is given.

Effects of Various Heating Rates on Glow Curves

R. Chen and S.A.A. Winer, *J. Appl. Phys.* **41** (1970) 5227.
This is a classic paper describing the theoretical basis of the heating rate methods for evaluating trapping parameters. Many details given in the derivation of the methods are very useful for everyone who wishes to study in depth the heating rate effects during TL measurements. The paper shows that Hoogenstraten's variable heating rate methods are valid for any general monotonically increasing heating rate function. The method of finding E by using the variation of maximum TL intensity I_m with heating rate is found to be applicable to all first-order TL peaks, and similar methods are introduced for general-order peaks. The heating rate method based on the peak maximum intensity I_m, surprisingly has not found much application up to now. Examples of applying the method of analysis to ZnS samples are given.

On the Calculation of Activation Energies and Frequency Factors from Glow Curves

R. Chen, *J. Appl. Phys.* **40** (1969) 570.
This is a classic paper that introduces several well-known peak shape methods of analysis of glow peaks. The equations are developed by using a combination of theoretical, empirical, and computational analysis for a wide variety of activation energies and frequency factors. Several formulas are developed for cases when the frequency factor depends on temperature, and for second-order glow peaks.

Glow Curves with General-Order Kinetics

R. Chen, *J. Electrochem. Soc.: Solid-State Sci.* **116/9** (1969) 1254.
Another classic paper where the peak shape method is developed for general-order kinetic peaks. The geometrical shape factor μ for general-order kinetics is calculated for values of the general order b between 0.7 and 2.5.

3. Papers Describing Kinetic Models in TL

A Critical Look at the Kinetic Models of Thermoluminescence: I. First-Order Kinetics

C.M. Sunta, W.E.F. Ayta, J.F.D. Chubaci, and S.Watanabe, *J. Phys. D: Appl. Phys.* **34** (2001) 2690.

The Quasi-Equilibrium Approximation and Its Validity for the Thermoluminescence of Inorganic Phosphors

C.M. Sunta, W.E.F. Ayta, R.N. Kulkarni, J.F.D. Chubaci and S. Watanabe, *J. Phys. D: Appl. Phys.* **32** (1999) 717.

These two papers examine the validity of the quasi-equilibrium (QE) assumptions commonly used in TL kinetic models. The papers include also a study of the conditions under which glow peaks of first-order kinetics are produced within TL kinetic models. Numerically computed glow curves without the QE approximations are calculated by using generalized multiple trap models. These glow curves are compared with analytically calculated glow curves, to verify whether the QE condition is satisfied. It is found that under a wide variety of combinations of parameters, the QE conditions are satisfied, even when retrapping is predominant over recombination. The paper concludes that the use of the QE approximation for analyzing glow curves is legitimate.

General Order and Mixed Order Fits of Thermoluminescence Glow Curves—a Comparison

C.M. Sunta, W.E.F. Ayta, J.F.D. Chubaci, and S. Watanabe, *Radiat. Meas.* **35** (2001) 47.

The authors compare the glow curves calculated using standard TL models with the glow curves obtained using general-order and mixed-order kinetics expressions. The goodness of fit is expressed by the figure of merit (FOM). They conclude that the mixed-order expressions characterize glow peaks more accurately than general-order expressions. They attribute this to the fact that the kinetic-order parameter b changes with temperature, while the mixed-order kinetics parameter α remains constant with temperature.

Theoretical Models of Thermoluminescence and Their Relevance in Experimental Work

C.M. Sunta, W.E. Feria Ayta, R.N. Kulkarni, T.M. Piters, and, S. Watanabe, *Radiat. Prot. Dosim.* **84** (1999) 25.

TL glow peaks are computed for the OTOR, NMTS, and IMTS models for a variety of input parameters. The characteristics of these calculated glow peaks are described, namely the effect of dose on the temperature of glow peak maximum, on the shape of the glow curve, and on the supralinearity of response. The results lead to the conclusions that the glow peak properties of the OTOR, NMTS, and GO models do not agree with the experimental properties of TL phosphors. The IMTS model on the other hand, is capable of producing glow peaks whose characteristics match with the experimental properties.

General Order Kinetics of Thermoluminescence—a Comparison with Physical Models

C.M. Sunta, R.N. Kulkarni, T.M. Piters, W.E.F. Ayta, and E. Watanabe, *J. Phys. D: Appl. Phys.* **31** (1998) 2074.

Pre-Exponential Factor in General-Order Kinetics of Thermoluminescence and Its Influence on Glow Curves

C.M. Sunta, W.E.F. Ayta, R.N. Kulkarni, R. Chen, and S. Watanabe, *Radiat. Prot. Dosim.* **71** (1997) 93.

In these two papers the authors study the behavior of the empirical parameters, the kinetic-order *b* and the pre-exponential factor *s*, which characterize general-order kinetics. Several TL glow curves are calculated within the OTOR, NMTS, and IMTS models and are analyzed using analytical methods, the shape of the glow curves and the isothermal characteristics. It is shown that *b* and *s* are not constant during the measurement of the TL glow curve, except when the kinetic order *b* is equal to 1 or 2. At the limit of very low trap occupancies (doses) the OTOR system produces second-order glow curves, and the IMTS model produces first-order glow curves. The implications of the general-order kinetics model for actual physical systems are discussed. The paper also shows that when appropriately defined, the pre-exponential factor also has a fixed value independent of trap occupancy. The empirical model seems to diverge from the experimental observations when the experimentally determined kinetics is non-first order.

Interactive Trap System Model and the Behavior of Thermoluminescence Glow Peaks

C.M. Sunta, W.E.F. Ayta, and S. Watanabe, *Mater. Sci. Forum* **239–241** (1997) 745.

In this paper the authors study a model consisting of a thermally active trap, a luminescence center, a deep thermally disconnected trap, and a shallow trap level. It is shown that such a model can explain several properties of experimental glow curves, like the shape, supralinearity properties of TL versus dose curves, sensitization by a predose, phototransfer, and stability of the peak positions.

General-Order Kinetics of Thermoluminescence and Its Physical Meaning

C.M. Sunta, W.E.F. Ayta, R.N. Kulkarni, T.M. Piters, and S. Watanabe, *J. Phys. D: Appl. Phys* **30** (1997) 1234.

This paper is an in-depth study of the empirical general-order kinetics (GOK) model, and an attempt is made to find a correlation between the empirical parameters b and s' in the GOK, with the physical parameters used in physically meaningful TL models. It is shown that the values of b and s' depend on the trap filling, and that the units of s' also change with dose.

On the General Order Kinetics of the Thermoluminescence Glow Peak

M.S. Rasheedy, *J.Phys.: Condens. Matter* **5** (1993) 633.
This paper introduces a new TL general-order equation in which the frequency factor is redefined in units of s^{-1}, and is also constant for a given sample and for a given constant initial trap concentration n_0 (dose). Nevertheless, this new frequency factor is found to vary when the sample dose n_0 is varied.

Mixed First and Second Order Kinetics in Thermally Stimulated Processes

R. Chen, N. Kristanpoller, Z. Davidson, and R. Visokecas, J. Luminescence **23** (1981) 293–303.
In this paper the mixed-order kinetics is shown to result from the more general set of three differential equations governing the "traffic" of carriers between a trap, the conduction band, and a recombination center under certain physical assumptions. Also, the applicability of this equation is discussed as an empirical approximation to the more general case. The solution of this equation is investigated and methods of experimentally extracting the trapping parameters of mixed-order kinetics are introduced. The advantages of the mixed-order kinetics presentation as opposed to the general-order kinetics models are discussed.

4. TL versus Dose Dependence Papers

On the Energy Conversion in Thermoluminescence Dosimetry Materials

A.J.J. Bos, *Radiat. Meas.* **33** (2001)737–744.
In the TL literature the thermoluminescence efficiency η of dosimetric materials is always taken equal to unity for the sake of simplicity. To the best of our knowledge, this is one of the few papers in the TL literature which looks at TL materials from the specific viewpoint of how efficiently they transform absorbed energy into easily detectable light (as a consequence to exposure to ionizing radiation). The maximum possible efficiency of well-known TL materials does not vary much and is found to be approximately 13%. Among the distinct steps in the conversion process (trapping, transfer, and recombination under the emission of light), the trapping appears to be the less efficient process.

Supralinearity and Sensitization of Thermoluminescence. I. A Theoretical Treatment Based on an Interactive Trap System

C.M. Sunta, E.M. Yoshimura, and, E. Okuno, *J. Phys. D: Appl. Phys.* **27** (1994) 852.

Supralinearity and Sensitization Factors in Thermoluminescence

C.M. Sunta, E.M. Yoshimura, and, E. Okuno, *Radiat. Meas.* **23** (1994) 655.
In these two theoretical papers the authors interpret the linear and supralinear behavior of TL versus dose curves within a model consisting of two electron traps and one recombination center. The model also provides an explanation and quantitative description for the predose sensitization observed in many TL materials. The case of LiF TLD-100 is used to demonstrate the applicability of the theory to actual experimental results.

Superlinearity in Thermoluminescence Revisited

R. Chen and G. Fogel, *Radiat. Prot. Dosim.* **47** (1993) 23.
In this paper a kinetic model consisting of two trapping states and one recombination center is presented. The model combines two previously published separate approaches based on the competition during excitation and on the competition during readout. The kinetic rate equations are solved without any simplifying assumptions and it is found that the model can explain the very strong superlinearity of the 110°C TL peak, which is observed experimentally in synthetic quartz.

5.5 eV Optical Absorption, Supralinearity and Sensitization of Thermoluminescence in LiF:Mg,Ti.

S.W.S. McKeever, *J. Appl. Phys.* **68**(2) (1990) 724–731.
The TL literature contains discussions of many competitive energy levels for dosimetric materials, but very few papers attempt to identify the nature of these competing centers. This paper is, indeed, the most serious attempt to identify the competitors responsible for the supralinearity of LiF:Mg,Ti. In the first part of the paper the author describes the properties that a competitor would possess. In the second part, the author discusses in detail the possibility that the well-known optical absorption band at 5.5 eV is the possible competitor.

Mechanism of Supralinearity in Lithium Fluorite Thermoluminescence Dosemeters

E.F. Mishe and S.W.S. McKeever, *Radiat. Prot. Dosim.* **29** (1989) 159–175.

This is a very fundamental paper which can be separated into two parts. The first part examines in detail the factors that affect the dose response function of LiF, such as linear energy transfer, impurity content, and heating rate. A comprehensive analysis of all data, lead the authors to the conclusion that the mechanism, which governs the TL dose response is operative in the heating stage of the TL process and not during the radiation absorption stage. In the second part a model for supralinearity is given, with a complete mathematical formulation, which successfully describes the observed experimental behavior. This paper is necessary for a deeper understanding of supralinearity as due to competition during the heating stage of TL.

Solution of the Kinetic Equations Governing Trap Filling. Consequences Concerning Dose Dependence and Dose-Rate Effects

R. Chen, S.W.S. McKeever, and S.A. Duranni, *Phys. Rev.* **B 24** (1981) 4931.
In this classic paper the authors solve the differential equations for the simple one-trap-one recombination center, one-trap-two centers and two-traps-one center models of TL. An additional period of time is introduced at the end of the excitation period, which allows the relaxation of the charge carriers in the bands. Results are obtained for various dose rates. The growth curves of TL versus dose are calculated and shown to yield superlinear behavior under appropriate choices of parameters.

Superlinear Filling of Traps in Crystals Due to Competition During Irradiation

S.G.E. Bowman and R. Chen, *J. Luminescence* **18/19** (1979) 345.
A simple model is studied consisting of two traps and one recombination center. The two traps are competing for electrons during the excitation period, leading to a linear-superlinear-linear-saturation behavior of the TL as a function of the dose.

Dose Dependence of Thermoluminescence Peaks

N. Kristianpoller, R. Chen, and M. Israeli, *J. Phys. D: Appl. Phys.* **7** (1974) 1063.
This is a theoretical investigation of the dependence of the maximum TL intensity I_M and of the corresponding peak temperature T_M on the excitation dose given to the sample. The model consists of an electron trap, a competing thermally disconnected deep trap and a recombination center. The kinetic equations are solved numerically and it is shown that superlinear behavior may arise within this

model, with the power of dose dependence equal to 2 or greater. The area under the glow peak is also studied as a function of the dose.

5. Review Papers in TL

Review: Models in Thermoluminescence

C. Furetta and G. Kitis, *J. Mater. Sci.* **39** (2004) 2277.

This recent review paper gives the fundamental equations and analytical solutions for several commonly used TL models, namely for the Randall–Wilkins, Garlick–Gibson, Adirovitch, May–Partridge, Braunlich–Scharman, Sweet and Urquhart, and mixed-order kinetics. The paper contains extensive results from general-order kinetics models and studies the influence of the properties of general-order peaks to dosimetry and to TL dating.

Luminescence Models

S.W.S. McKeever and R. Chen, *Radiat. Meas.* **27** (1997) 625.

This is an excellent review paper which can be very useful for new researchers in the field of TL modeling. It contains a description of the general one trap model (GOT), first-order Randall–Wilkins, second-order Garlick–Gibson, general-order kinetics, mixed-order kinetics, interactive kinetics, and the Schon-Klasens TL models. The quasi-equilibrium condition is examined and discussed extensively. Separate sections discuss tunneling phenomena and localized transitions. An extensive section presents several models that can describe different aspects of the growth of TL with dose: competition during excitation, competition during readout, and combined competition during both excitation and readout. Also contained in this paper are models for the optical bleaching of TL and for phototransferred TL. A review is given for several OSL models and the implications for dating techniques are discussed.

Kinetic Analysis of Thermoluminescence—Theoretical and Practical Aspects

Y. Kirsh, *phys. stat. sol. (a)* **129** (1992) 15.

This review article is organized in four sections: the first section contains the basic equations and results from several commonly used TL models. The second section reviews the main methods of analysis such as the initial rise method, curve fitting methods, peak shape equations for E, heating rate methods, and isothermal analysis. The third section discusses how these methods can be applied to complex TL curves and the last section presents additional experimental methods that can provide information about the TL process such as optical absorption, ESR, and TSC.

6. Papers on Curve Fitting and Deconvolution Functions

Fit of First Order Thermoluminescence Glow Peaks Using the Weibull Distribution Function

V. Pagonis, S.M. Mian, and G. Kitis, *Radiat. Prot. Dosim.* **93** (2001) 11–17.

Fit of Second Order Thermoluminescence Glow Peaks Using the Logistic Distribution Function

V. Pagonis and G. Kitis, *Radiat. Prot. Dosim.* **95** (2001) 225–229.

These two papers describe single glow peak algorithms which are available in several existing commercial programs. Analytical expressions are given which fit first- and second-order kinetics glow peaks. The proposed algorithms give excellent fits to TL glow peaks, although they are not physically based. Analytical expressions are given which allow an accurate evaluation of the activation energy E.

Thermoluminescence Glow Curve Deconvolution Functions for First, Second and General Order Kinetics

G. Kitis, J.M. Gomez-Ros, and J.W.N. Tuyn, *J. Phys. D: Appl. Phys.* **31** (1998) 2636.

Thermoluminescence Glow-Curve Deconvolution Functions for Mixed Order of Kinetics and Continuous Trap Distribution

G. Kitis and J.M. Gomez-Ros, *Nucl. Instrum. Methods Phys. Res. A* **440** (2000) 224.

In these two papers the authors develop several new analytical expressions for use in GCD analysis, several of which are also found in this book. The expressions describe accurately glow peaks following first- second- and general-order kinetics. Similar expressions are developed for mixed-order kinetics and for continuous trap distributions. The usefulness of these analytical expressions lies in the fact that two of the parameters I_M and T_M are determined experimentally. The accuracy of the expressions is tested by calculating the figure of merit (FOM) for synthetic glow curves.

Computerized Glow Curve Deconvolution: Application to Thermoluminescence Dosimetry

Y.S. Horowitz and D. Yossian, *Radiat. Prot. Dosimetry* (special Issue) **60** (1) 1995.

This special issue of the journal *Radiation Protection Dosimetry* covers almost everything in the TL literature regarding the method of glow curve deconvolution analysis. It is an absolutely necessary tool for everyone wishing to use glow curve analysis as a research and dosimetric analysis tool.

An Intercomparison of Glow Curve Analysis Computer Programs: I Synthetic Glow Curves

A.J.J. Bos, T.M. Piters, J.M. Gomez Ros, and A. Delgado, *Radiat. Prot. Dosim.* **47** (1993) 473.

An Intercomparison of Glow Curve Analysis Computer Programs: II Measured Glow Curves

A.J.J. Bos, T.M. Piters, J.M. Gomez Ros, and A. Delgado, *Radiat. Prot. Dosim.* **51** (1994) 257.

The series of these two papers presents the results of an evaluation of the capabilities of computer programs written to analyze glow curves in the framework of the GLOw Curve Analysis INtercomparison (GLOCANIN) project. The papers contain the results of an analysis of 13 different computer programs involving 11 participants from 10 countries on both computer generated and on experimentally measured glow curves. The intercomparison concentrated on the goodness of fit, the determination of the peak area, the temperature of the peak maxima, and the trapping parameters, i.e. activation energy and frequency factor.

7. Papers on Thermal Quenching and Temperature Lag Effects

Thermal Quenching of F-Center Luminescence in $Al_2O_3{:}C$

M.S. Akselrod, N. Agersnap Larsen, V. Whitley, and S.W.S. McKeever, *J. Appl. Phys.* **84** (1998) 3364.

Thermal quenching is an effect of importance in experimental thermoluminescence. This paper reports on experimental methods of evaluating the activation energy and frequency factor for thermal quenching. The paper also contains an analytical presentation of the heating rate method of TL glow curve analysis, and establishes that the quenching parameters are independent of sample type, degree of tap filling, cooling rate, and the heating rate. Finally, the influence of thermal quenching is simulated by numerical solution of the differential equations governing the processes.

Temperature Distribution in Thermoluminescence Experiments. I: Experimental Results

D.S. Betts, L. Couturier, A.H. Khayarat, B.J. Luff, and P.D. Townsend, *J. Phys D: Appl. Phys.* **26** (1993) 843.

Temperature Distribution in Thermoluminescence Experiments. II: Some Calculational Models

D.S. Betts and P.D. Townsend, *J. Phys D: Appl. Phys.* **26** (1993) 849.

Effects of Non-Ideal Heat Transfer on the Glow Curve in Thermoluminescence Experiments

T.M. Piters and A.J.J. Bos, *J. Phys. D: Appl. Phys.* **27** (1994) 1747.

A Simple Method to Correct for the Temperature Lag in the TL Glow Curve Measurements

G. Kitis and J.W.N. Tuyn, *J. Phys. D: Appl. Phys.* **31** (1998) 2065.
This series of four papers deals with the heat transfer effects inside the TL glow oven. The majority of TL readers use contact heating for the readout of the sample. The temperature recorded is the temperature of the thermocouple fixed on the heating strip, and not the temperature of the sample. However, when one wants to extract physical information from the glow curves it is essential to know the sample's true temperature. The above papers present experimental results concerning the estimation of the temperature difference between the heating strip and the samples called the temperature lag, and concerning the temperature differences between the lower and upper side of the sample, called the thermal gradient. They propose theoretical expressions for evaluating these effects and their influence on any physical information obtained from the glow curves, like the trapping parameters (activation energy and frequency factors).

Thermal Quenching and the Initial Rise Technique of Trap Depth Evaluation

S.A. Petrov and I.K. Bailiff, *J. Luminescence* **65** (1996) 289.
This paper describes the influence of thermal quenching on the activation energy values obtained with the initial rise technique. An analytical expression for correcting the activation energy obtained using this technique is given. Furthermore, the correction expression is generalized for any arbitrary form of internal thermal quenching behavior.

Effects of Cooling and Heating Rate on Trapping Parameters in LiF:Mg,Ti Crystals

A.J.J. Bos, R.N.M. Vijverberg, T.T. Piters, and S.W.S. McKeever, *J. Phys. D: Appl. Phys.* **25** (1992) 1249.

This paper is very good example of how trapping parameters (activation energy and frequency factor) are influenced by dynamic experimental parameters. These parameters include the cooling rate after a high temperature annealing, linear readout heating rate, and the use of a quadratic heating function. The discussion section contains an excellent presentation of the defect processes taking place during cooling rate, and readout heating rate, which influence the trapping parameters.

Thermal Quenching of Thermoluminescence in Quartz

A. Wintle, *Geophys. J.R. Astr. Soc.* **41** (1975) 107

This is the classic study of the effect of thermal quenching on the evaluation of the energy *E* using the initial rise method, for the 325°C thermoluminescence peak of quartz. The thermal quenching effect is confirmed by using radioluminescence measurements.

Appendix: A Brief Introduction to *Mathematica*

This appendix gives a brief introduction to some of the commands used in *Mathematica*. Only a very rudimentary listing of commands and examples are given here, and the reader is referred to the Mathematica Handbook.

The following short program uses the command **Plot** to graph the functions e^{-x^2} from $x = -3$ to $x = 3$.

```
Plot[Exp[-x^2],{x,-3,3}];
```

Mathematica produces the following output:

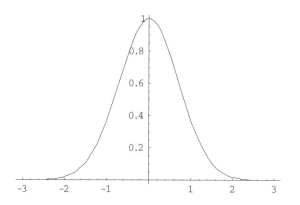

The following short program uses the command **Plot** to graph the functions x^2, x^3, and x^4 from $x = 0$ to $x = 1$. The graphs are stored in the graphic objects **gr1, gr2, and gr3**. Finally, the three graphs are shown together by using the *Mathematica* command **Show**.

```
gr1=Plot[x^2,{x,0,1}];
gr2=Plot[x^3,{x,0,1}];
gr3=Plot[x^4,{x,0,1}];
Show[{gr1,gr2,gr3}];
```

Mathematica produces the following output:

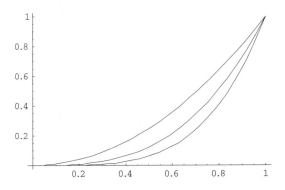

The following simple program in *Mathematica* solves the differential equation $y'(x) = ay(x) + 1$ with the initial condition $y(0) = 0$.

```
DSolve[{y'[x]==a y[x] + 1,y[0]==0},y[x],x]
```

Mathematica produces the following outut:

$$\left\{\left\{ y[x] \; \to \; \frac{-1 + e^{ax}}{a} \right\}\right\}.$$

The command **NDSolve** can be used to perform the numerical integration of the differential equations in the various TL models. For example, the following short program solves the differential equation $y'(x) = y(x)$ and stores the result of the numerical integration as the parameter **sol** (which stands for the **sol**ution of the differential equation). The integration is carried out from $x = 0$ to $x = 2$, and with the initial condition $y(0) = 1$.

The command **Plot** is used to graph the result of the numerical integration procedure from $x = 0$ to $x = 1$. The symbol "/.sol" in this program is interpreted as "given or using the values of the parameter **sol**."

```
sol=NDSolve[{y'[x]==y[x],y[0]==1},y,{x,0,2}];
Plot[y[x]/.sol,{x,0,1}];
```

Mathematica produces the following output:

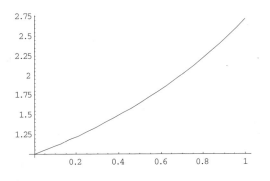

Mathematica uses simple useful objects called **Lists**. For example, the following is a list of (*x*, *y*) points called listXY. The points in the list can be graphed using the command **ListPlot**.

```
listXY={{0,0},{1,10},{3,40},{5,100}};
ListPlot[listXY];
```

Mathematica produces the following output:

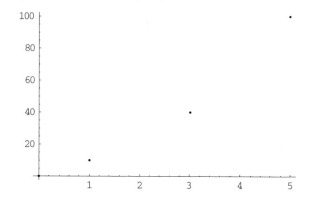

One method of producing a list of numbers is by using the command **Table**, which in the example below produces a list of numbers and their squares from $x = 1$ to $x = 10$ in steps of $x = 1.5$.

```
a=Table[{i,i^2},{i,1,10,1.5}]
ListPlot[a];
```

Mathematica produces the following output:

```
{{1,1},{2.5,6.25},{4.,16.},{5.5,30.25},{7.,49.},
    {8.5,72.25},{10.,100.}}.
```

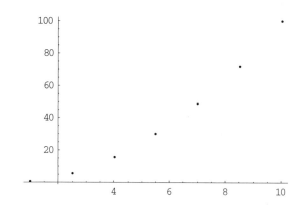

Mathematica can also solve systems of differential equations as seen in the following example which solves the system of equations $x'(t) = -y(t) - x(t)^2$ and $y'(t) = 2x(t) - y(t)$ with appropriate initial conditions. The command **Plot** is again used to graph the solutions $x(t)$ and $y(t)$.

```
sol=NDSolve[{x'[t]==-y[t]-x[t]^2,y'[t]==2x[t]-y[t],
    x[0]==y[0]==1},{x,y},{t,10}];
Plot[{x[t]/.sol,y[t]/.sol},{t,0,1}];
```

Mathematica produces the following output:

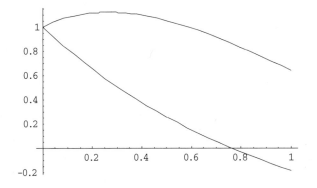

Author Index

Subject Index

A

activation energy, 1, 3, 7, 8, 10, 11, 12, 15, 17, 20, 24, 25, 26, 27, 28, 35, 36, 37, 40, 42, 44, 45, 47, 55, 58, 59, 60, 61, 62, 63, 70, 71, 72, 73, 76, 77, 80, 83, 97, 106, 109, 112, 115, 134, 144, 156, 161, 162, 164, 165, 166, 168, 171, 173, 174, 175, 177, 188, 196, 197, 198, 199
Al_2O_3, 156, 161, 197
annealing, 144, 147, 148, 149, 186, 189, 199
attempt-to-escapefrequency, 4

B

background, 24, 27, 70, 71, 72, 74, 147, 148, 149, 184, 185
band, 2, 83, 92, 94, 97, 98, 122, 123, 124, 125, 126, 128, 129, 132, 133, 134, 192, 193
Boltzmann constant, 37, 80, 83, 97, 109

C

calcite, 17
capture coefficients, 133
charge carriers, 194
charge conservation, 92, 96, 98, 123, 126, 129, 133
competition during excitation, 122, 128, 193
competition during heating, 122
competitor, thermally disconnected, 134, 193, 194
complex glow curves, 106
cooling cycle, 12
curve fitting, 20, 21, 25, 31, 32, 43, 49, 50, 59, 65, 106, 107, 109, 110, 182

D

differential equations, 79, 92, 93, 98, 120, 124, 125, 129, 130, 134, 135, 184, 192, 194, 197, 201, 203
disconnected, 96, 98, 134, 191, 194
dose rate, 155
dose response, 120, 121, 122, 130, 138, 185, 194
dosimeters, 138, 144, 145, 146, 147, 148, 149, 151, 152

E

E-Tstop, 11
electron–hole pairs, 124, 129, 134
environmental dosimetry, 144
exponential decay, 168, 169, 170, 187
exponential integral of TL, 109

F

first order kinetics, 87
FOM, 25, 32, 33, 34, 43, 50, 51, 52, 59, 67, 68, 79, 99, 101, 102, 103, 104, 105, 106, 107, 108, 109, 110, 111, 187, 190, 196
fractional glow (FG) method, 12
free electrons, 92, 94, 122
free holes, 122, 123, 129
frequency factor, 4, 7, 8, 10, 31, 33, 41, 47, 51, 55, 58, 67, 80, 83, 97, 106, 107, 109, 134, 165, 167, 188, 189, 192, 197, 199

G

Garlick–Gibson equation, 2, 195
GCD, 21, 106, 107, 196
general order kinetics, 4, 13, 80, 89, 91, 167
geometrical shape factor, 24, 28, 43, 45, 59, 61, 86, 91, 96, 158, 170, 189